ISBN 978-1-334-33564-8
PIBN 10653027

English
Français
Deutsche
Italiano
Español
Português

www.forgottenbooks.com

Mythology Photography **Fiction**
Fishing Christianity **Art** Cooking
Essays Buddhism Freemasonry
Medicine **Biology** Music **Ancient
Egypt** Evolution Carpentry Physics
Dance Geology **Mathematics** Fitness
Shakespeare **Folklore** Yoga Marketing
Confidence Immortality Biographies
Poetry **Psychology** Witchcraft
Electronics Chemistry History **Law**
Accounting **Philosophy** Anthropology
Alchemy Drama Quantum Mechanics
Atheism Sexual Health **Ancient History**
Entrepreneurship Languages Sport
Paleontology Needlework Islam
Metaphysics Investment Archaeology
Parenting Statistics Criminology
Motivational

SEMI-ANNUAL REPORT

OF

SCHIMMEL & Co.

(FRITZSCHE BROTHERS)

MILTITZ

NEAR LEIPZIG

LONDON ❧ NEW YORK.

———————⋈———————

OCTOBER/NOVEMBER 1904.

Contents.

List of abbreviations.

d = specific gravity at 15°, unless otherwise stated.

d_{200} = specific gravity at 20°.

$d\frac{20°}{4°}$ = specific gravity at 20°, compared with water at 4°.

$a_{D 150}$ = optical rotation at 15°, in a 100 mm. tube.

$[a]_D$ = specific rotation.

$n_{D 150}$ = index of refraction at 15°.

n = normal.

$\frac{n}{2}\left(\frac{n}{10}\right)$ = semi-normal or deci-normal (titrated solutions).

gm. = gram; cc. = cubic centimeter; mm. = millimeter.

Temperatures are uniformly stated in **centigrade degrees** (°).

Miltitz, near Leipzig, October/November 1904.

The opinion expressed in our April Report, that the East-Asiatic complications would have no pronounced effect on the economic conditions of Germany, has been confirmed by the course of events.

The Chemical Industry especially still continues to grow, and for the year 1903 again shows a gratifying expansion, which is reflected in the following figures. According to the statistics of the Association of Chemical Industry, the picture presents itself thus: —

	1903		1902
Number of Works	7747	against	7539
„ „ qualified Workmen	168950	„	160841
Amount of wages paid .	174402866,48 marks	„	164207621,06 marks,
of which must be taken into account	169335648,03 „	„	159655877,33 „

The export of products of the German Chemical Industry has grown in about the same proportion. It amounted to

448244000 marks	in	1903
430441000 „	„	1902
408366000 „	„	1901

Passing on to our branch of the Industry, the total exports of essential oils came to:

380000 kilos,	value	4634000 marks	in	1900,
388000 „	„	4658000 „	„	1901,
424700 „	„	5683000 „	„	1902,
424400 „	„	5942000 „	„	1903.

In contrast with this increase, the export of German perfumes and toilet-soaps unfortunately shows a falling off in value in 1903, whilst the quantities exported show an increase. This indicates a decline in the demand for expensive products, but, on the other hand, an advance in the demand for cheaper articles, — in other words, a fall in the average value of the exported products. The details are as follows:

I. Soap in tablets, including
    scented soap . . . . 4 806 500 kilos, value 6 008 000 marks
II. Fatty oils, scented . . . .   10 100   „   „   56 000 „
III. Scents, not containing alcohol   19 400   „   „   17 000 „
IV. Liquid perfumes, containing
    alcohol or ether, including
    hair-washes, dentifrices, and
    mouth-washes . . . . 1 717 400   „   „   5 158 000 „
V. All other perfumes, not other-
    wise enumerated . . . .   522 100   „   „   1 333 000 „
VI. Soap and perfumes insuffic-
    iently declared   900   „   „   5 000 „

Total in 1903 7 076 400 kilos, value 12 577 000 marks,

against in 1902, 15 570 000 marks
1901, 15 678 000 „
1900, 15 048 000 „
1899, 15 415 000 „

The principal falling off, amounting to nearly 2 000 000 marks, is under the heading IV. It is here not the place to enquire into the causes which have brought about this striking decline, but it is to be hoped that the difference will be equalised again in the current year.

With regard to the value of the exports of perfumes and common soaps from France, we have meanwhile found the following official data:

The exports were
in 1903, perfumes,   value 16 131 000 fr. against 14 699 000 fr. in 1902
„ 1903, soaps, not scented, „ 15 052 000 „   „   13 932 000 „ „ 1902

These figures, however, do not give a clear representation, as they do not appear to include perfumed toilet-soaps.

As regards the trade in our own branch, that with the United States has been very animated. It shows higher figures than for the same period of 1903. The situation has improved in so far that the price-cutting is no longer as keen as before.

The approaching Presidential election does not appear to influence business as much as used to be the case.

In the countries of Central and South America, the situation continues as changeable as must necessarily be the case in countries of a comparatively unstable condition. At this moment, Mexico and the Argentine Republic are progressing, whilst the state of affairs in Brazil, Chile and Venezuela is still uncertain.

In spite of the war, the export of our products to Japan was very brisk, and the pleasant relations with Russia also have up to now suffered in no way. In all the other civilised countries of Europe the demand for our products continued to increase.

The exceptional drought which prevailed throughout Central Europe during the spring and summer months was for the trade and industry in many respects most detrimental in its consequences, of which the stoppage of the river-traffic on the Elbe was the most serious of all. The loss caused thereby to the industry amounts to many millions of marks. The after-effects of the drought will make themselves still felt during the winter, and will find expression in high prices of food, spirits, and fodder.

For the development and growth of many medicinal plants and spices the lack of rain has been disastrous; our own cultivations have suffered so much from it, that the evil consequences will be felt for many years to come. The plain round Leipzig has been particularly affected by the drought. The districts of the Alps in the South of France have also suffered, and a considerably lower production is reported from there. In any case it will be well to be prepared for various surprises, which as a rule follow such natural phenomena.

The Northern parts of the American continent have also been affected by the abnormal drought, and it will be an interesting matter to determine to which exceptional metereological conditions these remarkable phenomena must be attributed.

As mentioned briefly in our last Report, we have taken up the manufacture of the hay-fever serum "Pollantin" (German Patent No. 152163), discovered by Professor Dunbar. Owing to the excellent action of this remedy, the demand was so large that unfortunately it could not be fully satisfied. As far as has been ascertained, the published results on the effect of this serum, with those of the present year added, are decidedly more favourable, and the demand next year will consequently be still larger. Arrangements have already been made to meet the increased requirements.

Almond Oil, bitter (from apricot-kernels). The certain knowledge of obtaining from us an absolutely pure natural distillate, has brought us a large circle of clients, whose demands we are scarcely able to satisfy in point of quantity. During the last few months it unfortunately happened that our stocks had been exhausted, and that some orders for larger parcels from abroad had to be declined. At the present moment we have, by working overtime, again accumulated a small stock.

On the occasion of some comprehensive experiments relating to the test of bitter almond oil for chlorine according to the well-known combustion-method, we established inter alia that the content of hydrocyanic acid in the oil may occasionally have a disturbing effect, and lead to errors. When abundant quantities of oil are used it may happen that the products of combustion carry along a small quantity of uncombusted hydrocyanic acid, and the latter is absorbed by the water present on the internal walls of the beaker. When the silver nitrate test is carried out, cloudiness will, under these conditions, occur even with pure oils, which cloudiness, however, is not caused by silver chloride, but by silver cyanide. We have repeatedly succeeded in detecting the hydrocyanic acid in such cases also by the Prussian blue reaction.

Contrary to silver chloride, the opalescence or cloudiness produced by silver cyanide disappears when the liquid is heated nearly up to boiling point, and a clear solution is obtained, whilst under the same conditions the cloudiness due to silver chloride remains. By these means we have succeeded in almost every case in differentiating the two compounds.

In order to avoid such errors, it cannot be too strongly recommended not to let the result of the examination depend upon only one single reaction, but to carry out a whole series of tests, — if need be alternately with a pure oil — and only from these draw a final conclusion. As the above-mentioned drawback does not by any means occur in every case, the total result will certainly not suffer in this manner.

It is further an advantage to proceed in this way[1]), that a piece of filterpaper, not too large (say about $2 \times 2^{1}/_{2}$ inches), folded in the manner of a pipe-light, is saturated with the oil, the excess of the latter thrown off by shaking the paper twice with the hand, the paper placed on a small porcelain dish and ignited, then immediately covered over with the beaker moistened on the inside, and the beaker left in that position for another minute after the flame is extinguished. The beaker is rinsed with 10 cc. water. In this manner we have always obtained reliable results, and we can therefore recommend the above-described method of working.

Almond Oil, pressed (from apricot-kernels). The apricot harvest in Syria in 1902 was a failure, though not to such an extent as in 1901. In consequence of the small stocks at the principal markets, the 1903 fruit was bought at comparatively high prices, which would probably have been driven up still higher if the Californian kernels had not prevented this.

The total production of Syria in the year 1903 is estimated at 4000 to 5000 bales. An interesting fact appearing in the report of the German Consulate at Damascus is, that whereas previously France was

---

[1]) For details compare also Gildemeister and Hoffmann, "The Volatile Oils", p. 440.

the only consumer of Syrian kernels, Germany now occupies the first place. Next come France and Italy, and with small shares Austria and the United Kingdom. Our own consumption amounts to about 250000 kilos per annum.

It is said that at Marseilles about 1200 bales kernels of previous years are still in the hands of speculators. These are useless for the expression of oil.

The result of this year's harvest has been normal. In consequence of a brisk demand and lack of stocks, it was not possible to depress the prices to any extent. On the contrary, the same price-factors as last year will have to be reckoned upon. The first consignment of this year's kernels is on the way for us.

Almond Oil, pressed (from sweet almonds). The price of genuine Pharmacopœia-oil has lately undergone no change, and it cannot be regulated until the prices of almonds of the new harvest have been fixed in Puglia and Sicily. From Bari, the world's principal market of almonds, it is reported that the present price is about 182 to 184 lire, and that the tendency is upward. The stock of old almonds is estimated at 20000 bales, the new harvest at 30000 to 35000. The total quantity is very small, especially if it is taken into consideration that a large proportion of owners will not sell before the new flowering-period in March/April next. The middlemen take all they can buy at above prices on speculation, and may possibly, up to the present, have bought up about 10000 bales. The exporters have entered into practically no engagements for autumn deliveries. The opinion is, that if Germany does not buy anything before the middle of September, the prices will fall for a time by about 10 to 12 lire, but that even when small orders are received, the quotations will again advance.

To say anything definite about the course of the market is impossible; it is probable that the selling prices of first quality ordinary sweet Bari almonds will be about 155 to 156 marks f. o. b. Bari.

Oil of Ambrette-seeds. For the liquid product there was a demand for kilo-parcels, which had the effect of clearing out the stocks. At the present moment a larger consignment of fresh Java seeds is in course of distillation.

Oil from Amorpha fructicosa L. The plant *Amorpha fructicosa* L., belonging to the *Galegeae*, a sub-family of the *Leguminosae*, which is frequently found in our gardens as an ornamental shrub, contains according to Vittorio Pavesi[1]), in its leaves and fruit various essential

---

[1]) Estratto dall' Annuario della Soc. Chimica di Milano **11** (1904), 4. Estratto dai "Rendiconti" del R. Ist. lomb. di sc. e lett. (II) **37** (1904), 487.

oils. On distillation with water-vapour the fruit yields 1,5 to 3,5 $^0/_{00}$, the leaves 0,5 to 0,8 $^0/_{00}$ essential oil of a bright colour and bitter taste which has the following constants:

fresh oil from leaves: $n_{D\,17,5°}$ 1,50083; $n_{D\,18,5°}$ 1,50928
old „ „ „ $n_{D\,17,5°}$ 1,50036; $n_{D\,18,5°}$ 1,50892.
Oil from unripe fruit: $d_{15°}$ 0,9019; $n_{D\,17,5°}$ 1,49951 } optical rotation feeble
„ „ ripe „ $d_{15°}$ 0,9055; $n_{D\,17,5°}$ 1,50036 } towards the left.

The portions of the leaf-oil boiling from 150° to 220° (750 mm. pressure), or respectively 80° to 120° (30 mm. pressure), contain a terpene which has not been identified further. In the higher boiling fraction, boiling point 250° to 265°, $d_{15°}$ 0,91661, $n_{D\,15°}$ 1,50559, cadinene could be detected by its hydrochloride melting at 117°. This fraction consists chiefly of a sesquiterpene, $d_{15°}$ 0,916, $n_{D\,15°}$ 1,50652, which approaches in its qualities clovene, or the companion of cadinene discovered by Wallach, or which may possibly be regarded as a new sesquiterpene for which the author proposes the name "amorphene".

Angelica Oil. Our own cultivations have suffered so much from the abnormal drought that only one quarter of an ordinary crop may be expected. The roots have but sparingly developed, owing to lack of nutriment, and are partly completely stunted. In Thuringia also the situation is not much better. In Belgium, on the other hand, the roots are said to have turned out well.

Anise Oil. Since the Russian anise-markets have taken place, it is now possible to survey to some extent the result of the harvest. In consequence of the cool weather which, contrary to the weather conditions in the Central European countries, has prevailed in the provinces where anise is indigenous, the harvest has been greatly delayed this year, and the quantities brought to the markets were smaller than usual. For these parcels very high prices were asked. The harvest is said to be unfavourable as regards quantity; the quality of the seed is good.

Our quotations of Russian anise oil may possibly be higher in the course of next month.

The cultivation of anise has recently received attention from the part of the Algerian Government. A supplement to No. 12 of the Bulletin de l'Office du Gouvernement général de l'Algérie (1904), contains a long article in which this subject is thoroughly discussed. The experiments made at Mustapha have succeeded extremely well. According to statements contained in this work, Spain produces about 2 000 000 kilos anise. The centres of production are 1. Province of Mancha: Daimiel, Manzanares, Membrilla, Bolanes, Torralva, Villaruba, et Quintanar de la Orden; 2. Province of Andalusia: Torrecampo,

Turredonzimeno, Villanueva de la Reina, Jamilena, Rube and Cazelle de la Sierra.

The production fluctuates between 480 and 600 kilos per acre.

France imported in 1902: —

from Spain	431 652	kilos
„ Russia	228 563	„
„ Turkey	630 402	„
„ other countries	84 068	„
Total	1 374 685	kilos,

having a value of 1 099 748 fr.

The cultivation of anise in Turkey appears to be continuously on the increase. According to a report from the German Consulate at Dedeagatch in Macedonia, the exports from that port in 1903 amounted to no less than 1 095 300 kilos anise, value 547 650 fr. Of this quantity, there were shipped

to France	865 100	kilos
„ Germany	170 000	„
„ Belgium	10 000	„
„ Austria	30 000	„
„ Italy	20 200	„
Total	1 095 300	kilos,

whilst in 1902 the anise-export of this port only came to 355 982 kilos.

In Thuringia and Moravia the anise-cultivation has practically become extinct. Chile, which country some 20 years ago sent us considerable quantities, now appears to consume its own production, for this article is no longer mentioned in the export-statistics.

Hoering [1]) makes a communication on peculiar reactions of some anethol derivatives. Is is a well-known fact that anethol dibromide reacts with sodium methylate with formation of anisyl-ethyl ketone; with monobromoanethol dibromide the same reaction is accomplished even more rapidly, not at all, however, with dibrominated anethol dibromide. This remarkable occurrence, and also the fact that he received from monobromoanethol dibromide (with one bromine atom in the benzene nucleus), by the action of concentrated nitric acid, a dibromoanisyl-bromoethyl ketone in which a second atom bromine had entered the benzene ring, led Hoering to agree with Semmler's [2]) view of the conjugated double-linkings of the side-chain and the benzene-ring. According to this view, a bromine atom in the α-side position

---

[1]) Berl. Berichte **37** (1904), 1542.
[2]) Berl. Berichte **36** (1903), 1033. Report October **1903,** 88 and April **1904,** 107.

(or also a methoxyl-group in the α-position) would under certain conditions enter into the nucleus:

$$CH_3O\!\!<\!\!\diagup\overline{}\diagdown\!\!>\!CHBr \cdot CHBr \cdot CH_3 \rightarrow \overset{\displaystyle HBr}{CH_3O\!\!<\!\!\diagup\overline{}\diagdown\!\!>}:CH \cdot CHBr \cdot CH_3.$$

During the further formation of the ketone dibrominated in the nucleus,

$$\overset{\displaystyle Br \quad H}{CH_3O\!\!<\!\!\diagup\overline{}\diagdown\!\!>}\underset{\displaystyle Br \quad H}{CO \cdot CHBr \cdot CH_3}$$

Hoering of course assumes a migration of the bromine atom entered in double, into the ortho-position towards the methoxyl-group, and supports this view by similar observations made by Zincke in the case of bromine additions in the eugenol series. Dibromanethol dibromide, whose abnormal behaviour also appeared in its indifference towards alcohol, yielded with sodium methylate, — contrary to the other dibromides — bromopropenyldibromanisol

$$C_6H_2\!\!<\!\!\begin{array}{l}Br_2\\OCH_3\\CH:CBr \cdot CH_3\end{array}$$

of the melting point $58°$. This body is not changed by bromine, but in the presence of hydrobromic acid as catalyser it absorbs atmospheric oxygen, and is converted with strong symptoms of luminiscence into a brominated ketone

$$C_6H_2\!\!<\!\!\begin{array}{l}Br_2\\OCH_3\\CO \cdot CHBr \cdot CH_3\end{array}$$

of the melting point $101°$.

In an experiment to substitute in monobromanisyl-bromoethyl ketone the bromine atom in the side-position by OH, $O \cdot COCH_3$, or $O \cdot COC_6H_5$, Hoering only obtained the two esters, but not the keto alcohol itself. He further obtained, as a by-product of the preparation of the tetrabrominated anethol (by dropping liquid anethol on bromine), in consequence of the splitting-up of the methoxyl-group by the hydrobromic acid generated, a tetrabrominated pseudo-phenol, which could also be produced from the tetrabromanethol itself by the action of the hydrobromic and glacial acetic acids. When this pseudo-phenol was boiled with glacial acetic acid and lead acetate, one of the bromine atoms in the side-position was substituted by acetyl.

According to Möhlau, E. Fischer, Bischler, and others, brominated ketones such as bromoacetophenone, β-bromolevulinic acid, etc., condense with primary aromatic amines into substituted indols:

$$C_6H_5 \cdot NH_2 + CH_2Br \cdot CO \cdot C_6H_5 = C_6H_4 \cdot NH \cdot CH = C \cdot C_6H_5 + HBr.$$

This reaction was pursued by Hell and Cohén[1]) with anisylbromoethyl ketone,

$$C_6H_4(OCH_3)CO \cdot CHBr \cdot CH_3$$

which they produced according to known directions from anethol dibromide. The brominated ketone, when heated with 2 molecules aniline in a tube to 125°, yielded a body, melting at 123°, of the formula

$$C_6H_4 \underset{\diagdown}{\overset{\diagup \text{NH} \diagdown}{\quad}} \overset{\diagup}{\underset{C \cdot CH_3}{\quad}} C \cdot C_6H_4 \cdot OCH_3$$

With o- and p-toluidine, and also with a-naphthylamine, the homologue indols were obtained under the same conditions. Hell and Bauer[2]) condensed the ethylisoeugenyl-bromoethyl ketone produced from ethylisoeugenol, with aniline and p-toluidine, and also obtained the corresponding indol derivatives. Acetylisation of the imide-hydrogen atom was found to be impossible. On oxidation there were formed anisic acid and ethylvanillinic acid respectively, with splitting up of the pyrrol-ring.

Oil from Artemisia herba alba. Emilien Grimal[3]) obtained from the fresh non-flowering herb of *Artemisia herba alba* Asso, a plant widely distributed in Algeria and much in favour there as a remedy, by means of steam-distillation, a yellow-greenish oil with a most agreeable odour. 150 kilos herb yielded 450 gm. oil, i. e. about 0,3 %. The oil which he examined further in our laboratory, had the following constants: $d_{15°}$ 0,9456; $n_{D20°}$ 1,47274; $(d)_{D20°}$ —15°38'; acid number 6,46; ester number 89,23; ester-content 31,15 %, calculated for $CH_3COOC_{10}H_{17}$, corresponding to 24,48 % alcohol $C_{10}H_{18}O$; ester number after acetylisation 135,38, from which, after deduction of the alcohol present as ester in the original oil, a content of 12,65 % free alcohol is calculated.

The oil readily dissolves in 2 to 2,5 parts 70 per cent. alcohol. When cooled to —12°, it did not solidify. On distillation in vacuo, fractions were obtained in which l-camphene, cineol and camphor could be detected. The higher boiling portions, when treated with

[1]) Berl. Berichte **37** (1904), 866.
[2]) Berl. Berichte **37** (1904), 872.
[3]) Bull. Soc. chim. III. **32** (1904), 694.

phthalic acid anhydride according to Haller's method[1]), yielded a small quantity of an alcohol which has not yet been examined further. From the saponification liquor of the oil,. after adding sulphuric acid, a fatty acid mixture was separated, which, according to the analysis of the silver salt, contained caprinic or caprylic acid.

Basil Oil. We have just received a small parcel of Réunion oil of quite exceptional quality. We understand that this distillate is used with advantage in mignonette-perfume compositions. It is extra-ordinarily rich.

In the essential oil of a species of basil (*Ocimum Basilicum* L.), P. van Romburgh[2]) has detected, in addition to considerable quanti-ties eugenol, a new olefinic terpene $C_{10}H_{16}$, ocimene. As an olefinic terpene, ocimene has a great resemblance to myrcene, the hitherto best-examined representative of this class of bodies. Like myrcene, ocimene is capable of absorbing oxygen, whereby it is converted into a colourless viscid substance. In its physical properties ocimene differs in a marked degree from myrcene, so that the two hydrocarbons cannot be regarded as identic. C. J. Enklaar, who has again taken up the examination of ocimene in the Utrecht laboratory, obtained by reduction with sodium in alcoholic solution, a dihydro ocimene which yields a crystallised bromine addition product, and which does not agree in specific gravity with the dihydro myrcene obtained by Semmler by reduction of myrcene.

Bay Oil. Two bay oils originating from Dominica, for which we are indebted to the kindness of Sir Daniel Morris, Imperial Commissioner of Agriculture in Barbados (W. I.), appear to point to the fact that the distillation there is not yet carried out with sufficient care, as the two oils were of abnormal character, and this can hardly be attributed to the Dominica bay-leaves, of which the excellent quality is well known.

The oils obtained from different manufacturers had the following properties: —

1. d_{15° 0,9500; phenol-content 55 %; soluble in 0,5 vol. 90 per cent. alcohol, cloudiness when more solvent was added.

2. d_{15° 1,0198; phenol-content 73 %; soluble in 0,3 vol. 90 and 95 per cent. alcohol, in both cases strong cloudiness when more alcohol was added. The dark-brown oil had the consistency of a viscid fatty oil, and differs considerably in point of odour from ordinary bay oil,

---

[1]) Compt. rend. **108** (1888), 1308; **122** (1896), 865.
[2]) Koninklyke Akademie van Wetenschappen te Amsterdam. Reprinted from Proceed. of the Meeting of Saturday, March 19, 1904. (April 19, 1904).

falling far behind the latter in fineness of aroma. The very low solubility, in spite of the high eugenol-content, shows that we have here
to do with an old distillate.

The anomaly between the two oils may possibly be due to the
fact that oil No. 1 consists only of the oil which in the distillation
separates off at the surface of the water, whereas in oil No. 2 the
portions of a bay oil distillate which have sunk to the bottom are
present. For the production of a normal bay oil the light and heavy
portions of the distillate must be combined.

Camphor Oil. The market of camphor oil, like all other Japanese
products, shows a certain nervous feeling which is an impediment for
the transaction of business. The advance which has taken place in
the camphor-price is associated with the repeatedly circulated rumours
that Japan would shortly be compelled to issue another foreign war-
loan, for which the camphor-monopoly would have to serve as security.
It is believed that in such case the Government would endeavour to
obtain control of the parcels of oil now held privately, and that the
article would only come on the open market again at decidedly higher
prices. This belief cannot be entirely rejected. Something must
undoubtedly take place, be it in this form, or in that of a tax on
the export, which will raise the price of this article, and we have
missed no opportunity of ensuring for a long time ahead our capability
of supplying camphor oil and its derivatives. For the time being we
are able to refrain from raising the prices of the latter.

By monopolising the production of camphor and camphor oil in
Japan and Formosa, the Japanese Government has obtained an influence
which is not exactly in favour of the consumers of these articles. For
this reason it is very desirable to create an effective competition
against the Japanese monopoly by the cultivation of camphor trees in
suitable countries, until perhaps some day the artificial camphor alone
takes this part upon itself. Germany is also interested in the question
of how to become independent of Japan as far as the supply of camphor
is concerned; and since in other tropical countries successful experiments
have been made in planting *Cinnamomum camphora* Nees & Eberm., a
commencement has now also been made in the German colonies with
the cultivation of camphor-trees, as is reported in communications
from the Biologico-agricultural Institute at Amani[1]).

The camphor-trees are best grown from seed, which is gathered
in Japan in the autumn, and which after being dried, is kept during the

---

[1]) Reprint from the "Usambara Post". Communications from the Biologico-
agricultural Institute at Amani, January 9, 1904, No. 9.

winter in white sand. According to Bamber and Willis, the seed, before sowing out, is soaked in a suitable manner for 24 to 48 hours in water, and then sown at a depth of $^1/_2$ to $^3/_4$ inch. When the plants have attained a height of 10 to 16 inches, they can be planted out for good. The space between the plants varies, according to whether it is desired to obtain the camphor from the leaves or from the wood of the trees; in the former case it may be convenient to plant in the form of hedges, but in the latter, to leave more room between the plants, — say 15 or 16 feet in every direction. The camphor-tree can be propagated not only from seed, but also by layers and shoots. Experiments in this direction will also be made by the Biologico-agricultural Institute.

Of *Dryobalanops aromatica* Gärtn., which yields the Borneo camphor, there are up to the present only a few young plants in the experimental garden at Amani. As soon as an opportunity presents itself, cultivation experiments will be made on a larger scale.

According to the information from Amani, it is, however, not considered advisable to plant the composite *Blumea balsamifera* D. C. (which yields the Blumea camphor) in the German Colony.

Cananga Oil. In the course of the summer we have received a large consignment of several thousand bottles direct from Batavia, and after having rectified the oil in vacuo, we now offer it for sale. The quality of this parcel is beautiful. The price is calculated as low as possible, but is naturally higher than that of the crude commercial oil which contains a large quantity of resin and impurities, and which on rectification shows a considerable loss.

Caraway Oil. The abnormal heat and drought have also influenced the Dutch caraway-harvest. Although the yield per acre is on the whole a normal one, the herb of the crop in Zeeland and Brabant, which was cut somewhat too early, is for the greater part stunted, and smaller than usual. But in other districts, and especially in those from which during the last decades we have chiefly drawn our supplies, the fruit is well developed, and shows an abundant content of oil.

The area planted with caraway has remained stationary during the last few years, and will probably remain so. Whereas previously caraway was cultivated exclusively in North Holland, it has in more recent times found its way to South Holland, but it did not thrive there (probably owing to unsuitable soil) and the cultivation was soon given up. But in the province of Zeeland, where the former cultivation of madder has been replaced by caraway, good results have been obtained; the same is the case in Brabant and Flanders, where many farmers now grow caraway instead of sugar-beet.

The area cultivated with caraway is estimated at about 8750 acres, divided as follows: —

North Holland . . .	about	4 000	acres
Brabant and Flanders	„	2 250	„
Zeeland	„	2 000	„
Friesland	„	375	„
South Holland . . .	„	125	„

Total about 8750 acres.

Our confidential informant, who has grown grey in the caraway-trade, estimates the results of this year's harvest as follows: —

in North Holland . .	about	48 000	bales
„ Zeeland	„	23 000	„
„ Brabant and Flanders	„	25 000	„
„ Friesland	„	4 000	„
„ South Holland . .	„	1 000	„

Total about 101 000 bales,
to which must be added
Stocks remaining from former years about   40 000 „

Grand total 141 000 bales.

Among the old stocks there is a good deal of inferior stuff, spoilt by rain, which must be disposed of. Under these conditions an advance in the value of this article can scarcely be expected, although the present market quotation of fl. $11^1/_2$ to fl. 12,— leaves hardly any profit to the grower, and is not remunerative, as compared with other produce of the field.

The prices of caraway oil and carvol are calculated on our purchases to date; they come a trifle lower.

From other caraway-producing countries, no reliable reports have as yet been received.

Cassia Oil. The price-fluctuations during the last six months have been of a very unimportant character. The reports from China have now for some time been absolutely silent about this article.

Our quotations are worthy of notice.

Cassie „Schimmel & Co.“ (Artificial oil of cassie-blossoms, German Patent No. 109635 and additional Patent No. 150170). Since the publication of our last Report we have added this product to our lists as a novelty.

As our various publications[1] since October 1899 show, we have paid special attention to the study of the chemical constituents of the

---

[1] Reports October **1899**, 58; April **1901**, 18; April **1903**, 17; October **1903**, 18; April **1904**, 23; Journ. f. prakt. Chemie **68** (1903), 235.

natural cassie oils from the blossoms of *Acacia Farnesiana* Willd. and *Acacia Cavenia* Hook., and have detected in these oils a whole series of bodies, the use of which has been protected by patents.

The product which as a practical result of these researches has been introduced by us into commerce, is a mixture of these constituents with natural blossom oil. It contains no diluent whatever, and is consequently so extraordinarily rich, that 5 gms. of it dissolved in 1 kilo strongest alcohol, form an extrait triple of excellent strength and fine natural floral odour.

We can justly describe the effect of our artificial cassie oil as excellent, and can recommend it as a first-rate product which is called to render good services to the perfumery-industry.

We shall be glad if clients will verify for themselves the truth of the foregoing statements by trials, and will show the same interest in this new product, as is so largely the case with our other blossom-oils which in point of quality occupy the highest possible position.

Cedarwood Oil. The waste product of fine Florida cedarwood is becoming more and more scarce, as the manufacturers of lead-pencils now mostly order the wood sawn in the form of small boards, which yield very little sawdust; for this reason higher prices must now be paid.

In view of the growing consumption of cedarwood, it is not impossible that it may be necessary to obtain supplies of the wood in logs, in which case the value of the article would undergo a further increase. We would not fail to point to the possibility of such a course.

Chamomile Oil. Our stocks consist exclusively of freshly distilled oil of highly aromatic Hungarian chamomile. This oil has a beautiful intensely blue colour.

Cinnamon Oil, Ceylon. The price of fine cinnamon-chips remains at the low level of 2 d. The cheap quotations of our finest, heavy, sweet distillate could therefore be maintained without change.

Under these circumstances, the chances of the artificial Ceylon cinnamon oil are not favourable.

Citronella Oil, Ceylon. Since the date of our last Report, the prices in Colombo have gone higher. Business remained dull, larger parcels being rarely offered. From this it may be gathered that the bulk of the production has been sold for forward delivery.

The shipments which in 1903 showed a decline of more than 200000 lbs. as compared with 1902, now again show an increase; to the end of July, i. e. in 7 months, they came to

710071 lbs.

against 569875 „ to the end of July 1903
„ 725953 „ „ „ „ „ „ 1902
„ 681676 „ „ „ „ „ „ 1901

Of the above quantity

300847 lbs.	went	to	the	United Kingdom
325419 „	„	„	„	United States
37459 „	„	„		Germany
35452 „	„	„		Australia
6528 „	„	„		China
3681 „	„	„		India
270 „	„	„		Russia
235 „	„	„		France
180 „	„	„		Belgium
710071 lbs.				

The statistics are no criterion for the trade with the individual countries, as the quantity exported to the United Kingdom includes important consignments for our firm, which are shipped via London to Hamburg. Equally important as our consumption here, is that of our New York house. The two together absorb fully one third of the whole production.

Large quantities of citronella oil are used up in the manufacture of geraniol, which with us is constantly increasing.

With regard to the value of this article, a decline is improbable, but a rise is not out of the question in face of the present situation. We do not see any risk in contracts at the present quotations.

The Government of Ceylon now appears to be prepared to take energetic measures for counteracting the decline of the citronella oil industry there which would be the natural consequence of the extensive adulterations, and to create again a healthy situation, as the powerful development of the Java oil industry also forms a threatening danger.

A communication from the Government[1] to the Chamber of Commerce at Colombo shows, that the intention is to accept the proposal made by J. C. Willis, Director of the Royal Botanical Garden at Peradeniya, according to which the citronella oils for export are to be subject to strict supervision. For this purpose, samples are taken from each oil drum, previous to shipment, by specially appointed inspectors in the presence of a representative of the export firm, and these samples tested immediately according to Bamber's method. A certificate of the result of the examination is handed to the firm, and the drums are officially sealed. The oils which according to Bamber's test do not contain more than $1^0/_0$ of adulterations are sealed with a red seal which, in addition to the Government brand, carries the remark *"pure oil"*, whilst oils which are found to be adulterated

---

[1] Reprinted in "The Ceylon Observer" of August 16, 1904.

up to $10^0/_0$, are marked with a green seal with the words *"90 per cent. purity"*. The export of all oils adulterated to a greater extent is prohibited.

The export firms have the right to appeal against the results declared by the inspector; in such case a further sample is taken and sent to Bamber; the result of his examination is final. For this test certain fees are payable, which are refunded if an error is proved against the inspector.

Such stations for examining the oil would be established at Galle and Colombo, the two export ports which come exclusively under consideration for citronella oil. The inspectors must of course have sufficient experience for carrying out the tests, and be thoroughly reliable. They should be placed under the supervision of a higher official, who would have the power of removing the seals at any time and taking a second sample from the drums, which would then be tested by Bamber. A check of this kind would have to be made frequently and unexpectedly.

The expenses in connection with this new arrangement might be met by a corresponding export tax.

It is to be hoped that the Government measures indicated here will now also become an established fact. This would no doubt meet with general approval, in the interest of the honest trade in Ceylon citronella oil. But in view of our present experience, we doubt whether Bamber's method will be found suitable for the examination in question.

In our previous Report (p. 29 et seq.) we have already stated that Bamber's method is useful for the qualitative test for adulterations, although in our experience it is not free from the risk that unadulterated oils will also frequently be objected to, because they may not pass Bamber's test; further down we refer again to such an oil. But it appears to us risky that it is also intended to indicate the adulteration quantitatively, as this method, as shown by us at the time, does not give reliable results in this respect. It may be anticipated that this will very often give rise to differences of opinion.

As the real value of Bamber's method would lie in the quantitative estimation of the adulterant, but as it does not give reliable results, we scarcely believe that this method is to be preferred to the "raised Schimmel's test"[1]) recommended by us, which latter consists of this, that the citronella oil in question, after the addition of $5\,^0/_0$ Russian petroleum, is tested for its solubility in 80 per cent. alcohol. A pure citronella oil should under these conditions dissolve in 1 to 2 vol. 80 per cent. alcohol, and even when as much as 10 vol.

---

[1]) Compare Report April **1904**, 32.

of the solvent is added, at most give an opalescent cloudy solution; separation of oil should not take place. The observation-temperature is $15°$ to $20°$. This method gives very reliable results, and has also been employed with advantage in a similar manner in Ceylon. With regard to simplicity in the execution, it is undoubtedly superior to Bamber's method.

Fortunately the situation in Ceylon appears in other respects to be gradually improving; even the producers seem to understand that a continuation of the present system means certain ruin to the citronella oil industry. Good oils are now frequently placed on the market, and we ourselves have quite recently received oils of such excellent quality as we have not had for a considerable time. The content of total geraniol (geraniol $+$ citronellal) amounted to about $60°/_0$. These oils dissolved well in 80 per cent. alcohol, and also passed the above-mentioned „raised Schimmel's test." The dilute alcoholic solution of the oil mixed with $5°/_0$ Russian petroleum no doubt showed a strong opalescence, but no oil was separated off even at a low temperature (refrigerator). In spite of this, a test carried out by Bamber's method showed an adulteration of about $5°/_0$, although all other indications of such adulteration werea bsent. In our opinion the oil examined is pure, and of faultless quality, and we consider that the result obtained by Bamber's process shows a defect in that method.

Citronella Oil, Java. The manufacturers in Java have by larger shipments carried out their promise to increase the production, and it has therefore up to the present been possible to execute in full all the orders which have been received. It is to be hoped that this business, now so well arranged and developed, will escape the natural phenomena which are of such frequent occurrence in Java.

Clove Oil. Important fluctuations, due chiefly to speculative causes, have again to be recorded in the prices of Zanzibar cloves since we published our last Report. The highest level was reached in May with fl. 46,— per 50 kilos, the lowest in July with fl. 36,—, for prompt delivery. At present (the middle of September the quotations on the Dutch er minal market are as follows: —

October	1904 . . .	fl.	$34^8/_8$
November	„ . . .	„	$32^5/_8$
December	„ . . .	„	$31^3/_4$
January	1905 }		
February	„ }	. .	„ $31^7/_8$
March	„ . . .	„	$30^7/_8$
April/May	„ . . .	„	$31^8/_8$

The result of the new harvest is purposely kept secret, or else such contradictory reports are given out about it that it is impossible to form a clear judgment. The uncertainty is at present so great that we cannot make up our mind to urge clients to buy. As a matter of fact, we consider the upward tendency an artificial move of the speculative interest. The new harvest is said to give an abundant yield.

The statistics of the quotations of the last 6 months show the following fluctuations: —

beginning of April	fl. $39^1/_2$	end of April	fl. $41^1/_2$
„ „ May	„ 44	„ „ May	„ 46
„ „ June	„ $44^1/_2$	„ „ June	„ $42^1/_2$
„ „ July	„ $40^3/_8$	„ „ July	„ $40^1/_8$
„ „ August	„ 40	„ „ August	„ $40^1/_2$
„ „ September	„ 36	„ „ September	„ 34
„ „ October	„ 34	„ „ October	„ 32

When the price of Zanzibar cloves had risen to about fl. 40,—, Amboina cloves, of which large stocks had accumulated in Holland, were also drawn in on the Amsterdam terminal market, and so-called "mixed contracts" were introduced, which could be carried out half in Zanzibar and half in Amboina cloves. The latter have a large oil-content and yield a distillate of excellent aroma.

An oil obtained by us from powdered Amboina cloves had the following constants: $d_{15°}$ 1,0456; α_D —$1°24'$; phenol-content 78 to 79 $\%$. The clear solution of the oil in about an equal volume 70 per cent. alcohol, showed a fairly strong turbidity with 3 to 4 vol. of the solvent, remaining when more alcohol was added. We would call special attention to this, as in the case of ordinary oil of cloves even the dilute solution in 70 per cent. alcohol is clear. The advantage of the Amboina oil consists in the specially fine odour, which renders it decidedly superior to the ordinary distillate. This fact may ensure a permanent place in perfumery to Amboina clove oil.

Coriander Oil. For purposes of distillation, the Russian, Moravian, and Hungarian seeds come this year under consideration. Thuringia reports a failure in the harvest. The prices of the oil are high.

Cypress Oil. In consequence of the information given in our last Report on the excellent results obtained with this oil in whooping-cough, such a large demand has arisen that the stocks in hand were soon used up, and the sale had to be suspended completely for several weeks, as it was not possible to provide at that time of the year the material for distillation. A few weeks ago the first shipments have again arrived, and they have immediately been taken in hand, so that now, and we hope also for the future, this remedy can be supplied in a more liberal manner.

We have again received reports on the excellent effect of the oil from several physicians who express themselves with extraordinary appreciation. The number of paroxysms of whooping-cough was reduced in a few days by one half and even more. A practitioner at Friedenau who is very sceptical about whooping-cough remedies, observed in the case of a child which had already been ill for three weeks, that the paroxysms diminished from 14 to 18, down to 8 to 10 daily, and with two younger children, 2 to 4 years old, the number of attacks fell within a week from 35 to 14, and 41 to 18 respectively. This physician calls cypress oil a very prominent remedy against whooping-cough, and praises its extraordinary rapid action in wholly different stages of the disease.

Owing to the brisk demand we had opportunities to submit various distillates to our research-laboratory. In one of our own distillates the ready solubility was very striking, whilst a fresh French distillate, which otherwise was entirely unobjectionable, was characterised by a low specific gravity and low ester- and alcohol-content. The properties of our own distillate were:

$d_{15°}$ 0,8916; a_D $+16°27'$; acid number 1,88; ester number 19,53; ester number after acetylisation 48,48; the oil dissolved in 2,5 vol. 90 per cent. alcohol. The examination of the French distillate gave the following values: $d_{15°}$ 0,8680; a_D $+26°31'$; ester number 5,31; ester number after acetylisation 10,25; the oil was soluble in 5,5 and more vol. 90 per cent. alcohol.

After having further examined the chemical composition of cypress oil[1]), we are now in a position to give the following information on this subject.

In the high-boiling portions a sesquiterpene, l-cadinene, was found in addition to cypress camphor. For the purpose of isolating the sesquiterpene, we repeatedly fractionated in vacuo the portion of the last runnings of a cypress oil of the boiling point 120° to 137° (5 mm. pressure), which had been freed from camphor by freezing out and stirring up with dilute alcohol, and we purified the fraction boiling at 130° to 135° from oxygenated admixtures by repeated, prolonged heating with metallic sodium. In this manner we obtained a colourless oil, whose odour reminded of cadinene, and which had the following constants: boiling point 270° to 272°; $d_{15°}$ 0,9203; a_D $-4°35'$; $n_{D20°}$ 1,50621. The dihydrochloride formed from it with gaseous hydrochloric acid, with ice-cooling, melted after recrystallisation from acetic ether at 117° to 118°, and was thereby recognised as cadinene hydrochloride. Its specific rotation determined in 10,0 per cent. chloroform solution was — 25° 10'.

---

[1]) Report April **1904**, 38.

We next turned our attention to the already known cypress camphor which forms an important constituent of the oil. It is chiefly present in the portions distilling above 135° (5 mm. pressure), from which it crystallises out in long needles, partly spontane-ously and partly by application of the agents mentioned above. In consequence of its great capability of crystallising, which is a property of most "camphors", its production in the pure state does not offer any further difficulties. At the same time, it has to be recrystallised 5 to 6 times from dilute alcohol, and to be purified twice from petroleum ether, in order to obtain the compound absolutely odourless. With regard to the chemical nature of this body, it had up to now been taken for a sesquiterpene alcohol. With a view to deciding this question according to its constitution, we have analysed it, determined its molecular weight, and tested it for its behaviour towards water-abstracting agents, and we have thereby found that it actually is a sesquiterpene alcohol of the formula $C_{15}H_{26}O$. The product worked up by us was odourless, optically inactive, and melted readily at 86° to 87°, that is to say 1° higher than according to previous statements[1]). It distilled not very constant at 290° at 292°.

0,1635 gm. of the substance: 0,4863 gm. CO_2, 0,1726 gm. H_2O.

Found:	Calculated for $C_{15}H_{26}O$:
C 81,11$^0/_0$	81,08$^0/_0$
H 11,73$^0/_0$	11,71$^0/_0$.

The molecular weight, determined according to the boiling method of Beckmann in benzene solution, was found to be 210,7 and 235,3 (calculated 222).

Strong formic acid converted the body with loss of water quantitatively in a hydrocarbon $C_{15}H_{24}$. For this purpose we shook 50 gm. of the camphor with three times the quantity of 99 to 100 per cent. formic acid until solid camphor was no longer present. The hydrocarbon was driven over with water-vapour and distilled over sodium. Its physical constants were as follows:

boiling point 264°; $d_{15°}$ 0,9367; $[a]_D +94°3'$; $n_{D20°}$ 1,49798.

Elementary analysis showed that the product is a hydrocarbon $C_{15}H_{24}$.

0,1470 gm. of the substance: 0,4731 gm. CO_2, 0,1552 gm. H_2O.

Found	Calculated for $C_{15}H_{24}$
C 87,78 $^0/_0$	88,23 $^0/_0$
H 11,73 $^0/_0$	11,76 $^0/_0$.

---

[1]) Report October **1897**, 12, note. Comp. Gildemeister and Hoffmann, "The Volatile Oils", page 269.

Attempts to produce from the sesquiterpene the above alcohol, a dihydrochloride, nitrite, or nitrosate, had only negative results. We were, however, successful in obtaining a nitrosochloride, but this crystallised with exceptional difficulty owing to resinous impurities. Only once we obtainèd a few crystals by diluting the crude product with a trace of alcohol, and then cooling it very strongly. All further attempts remained unsuccessful. The hard crystals referred to melted indistinctly at 100° to 102°.

In connection with these results, we were desirous of deciding whether the cedar camphor or cedrol present in cedar oil is actually, as has hitherto been assumed[1]), the optically active modification of the inactive cypress camphor, or whether it represents another sesquiterpene alcohol. For this reason we determined its physical constants, which, apart from the optical activity, entirely corresponded with those of cypress camphor. Its analysis, determination of molecular weight, and also its behaviour towards formic acid, also led to exactly the same results as in cypress camphor. The analytical data follow here:

melting point 86° to 87°; boiling point 291° to 294°; $[\alpha]_D + 9° 31'$
(determined in 11,2 per cent. chloroform solution).

0,1593 gm. of the substance: 0,4735 gm. CO_2, 0,1685 gm. H_2O.

	Found	Calculated for $C_{15}H_{26}O$
C	81,06 %	81,08 %
H	11,75 %	11,71 %.

Molecular weight (according to Beckmann's boiling method)

Found	Calculated
249	222

Hydrocarbon:
boiling point 263,5° to 264°; $d_{15°}$ 0,9366; $\alpha_D - 85° 32'$;
$n_{D20°}$ 1,49817.

0,1359 gm. of the substance: 0,4393 gm. CO_2, 0,1466 gm. H_2O.

	Found	Calculated for $C_{15}H_{24}$
C	88,16 %	88,23 %
H	11,99 %	11,76 %.

In this case also, all attempts to produce a crystalline derivative of the hydrocarbon were unsuccessful. In the nitrosochloride test, however, the course of a reaction could here also be observed. The difference between the two bodies therefore only consists of this, that cedar camphor is dextrorotatory, and cypress camphor inactive, and that the former under the action of strong formic acid is converted

---

[1]) Gildemeister and Hoffmann, "The Volatile Oils", loc. cit.

into a lævorotatory, and the latter into the same dextrorotatory hydrocarbon. In their other physical and chemical properties the two are identic. This is also proved by the fact that the mixture of cypress camphor and cedar camphor shows no depression of the melting point, but melts uniformly at 86° to 87°.

The distillation residue of cypress oil forms a viscid brown resin with a ladanum-like odour. We have also occupied ourselves with the examination of this product which is valuable for the odour of the oil, but up to the present we have not arrived at any particular result.

We may here mention that the assumed terpene alcohol[1]) contained in the fractions boiling at 90° to 95° (4 mm. pressure), contrary to our previous observation[2]), does form a phenyl urethane of the melting point 142° to 144°.

As the result of our examination we have detected in cypress oil: furfurol, d-pinene, d-camphene, d-sylvestrene, cymene, a ketone, sabinol (?), a terpene alcohol (?), d-terpineol of the melting point 35° as ester (probably as acetate), valerianic acid, l-cadinene, a sesquiterpene alcohol, cypress camphor (identic with the sesquiterpene alcohol of cedar oil), and a body with a ladanum-like odour. Of these, d-pinene and also cypress camphor had already previously been detected by us in the oil.

Elemi Oil, Tacamahaca. During their further researches on secretions, A. Tschirch and O. Saal[3]) isolated by steam distillation from Tacamahaca elemi, an essential oil in a yield of 2%. It had a bright yellow colour and a peculiar aromatic odour reminding more of borneol than of the oils obtained from elemi. On fractional distillation of the oil, a colourless oil with a pleasant odour passed over at 70°, whilst the bulk distilled over at 170° to 175° in the form of a bright-yellow oil with a faint empyreumatic odour which became more pronounced in the following fractions (190° to 195° and 220°). Above 220° there passed over a dark-yellow to brown oil possessing an unpleasant pungent odour. A thick, dark-brown mass with a strong tarlike odour remained behind as residue.

The same authors[4]) obtained in the same manner from genuine commercial Tacamahaca elemi about 3% of a yellow essential oil whose odour reminded very distantly of that of the typical elemi resins, but more of camphor and turpentine. The bulk of the oil distilled over between 170° and 175° as a colourless oil, whilst between 175° and 210° the remaining oil passed over as a dark-coloured portion which possessed a somewhat empyreumatic odour.

---

[1]) Report April **1904**, 40.
[2]) loc. cit.
[3]) Arch. der Pharm. **242** (1904), 362.
[4]) Arch. der Pharm. **242** (1904), 400.

Essential Oils, Sicilian and Calabrian.

With regard to this important group of essential oils we have received the following original report from the well-known reliable source: —

The first 8 months of the current year show a considerable increase in the export of essential oils, as compared with the same period of last year, and reach with 612 737 kilos almost the figures of the export of 1902, which was quite exceptionally high.

The following summary gives the export figures of the individual months, in comparison with the corresponding months of the two preceding years: —

	1904	1903	1902
January . . .	105 877	95 975	132 509 kilos
February . . .	98 897	97 646	74 056 „
March . . .	91 132	78 390	113 977 „
April	74 955	67 319	83 453 „
May	57 932	62 452	77 291 „
June	84 286	36 404	45 059 „
July	52 584	52 176	61 247 „
August . . .	47 074	46 340	52 567 „
	612 737	536 702	640 159 kilos

The view repeatedly expressed in these Reports, that the world's consumption is constantly increasing, appears to be confirmed by the above figures, in spite of the decrease from 1902 to 1903.

Thus far the total picture of the essential oil trade. Of the course of the business in the individual kinds of oil, the following lines may give an idea.

Bergamot Oil. In spite of the surprisingly brisk demand, which prevailed during the first few months of the manufacturing season and which gave rise to the opinion that during the spring large stocks of bergamot oil would accumulate abroad, a strong demand for this oil made itself felt in May/June, and considerable quantities were shipped in the course of these two months.

As at the same time the weather, which up to then had been favourable for the bergamot-trees and had promoted an abundance of blossoms, exerted a very unfavourable influence in consequence of the extraordinary early heat which was by no means tempered by short showers, and caused enormous masses of tender fruit-germs to fall off the trees, a sharp rise in the prices was bound to follow. At the beginning of May the price stood at 16 marks, but it advanced in the course of the next few months and early in July to 19,75 marks, and in some cases even to 20,50 marks; but when the buyers abroad curtailed their purchases, and the demand diminished considerably, the

Calabrian speculators were compelled to become gradually less exacting, although the prospects of the coming harvest were by no means favourable.

The total oil-harvest of last year may, as a matter of fact, be estimated at about 100000 kilos, of which at the moment of writing perhaps 8000 or 10000 kilos are still in the country; but the estimates of the coming harvest come only to about 40000 to 50000 kilos, provided the fruit gives a normal yield of oil. Under these conditions a higher basis of prices than last year is to be expected from the outset for the coming oil-producing season.

Lemon Oil. The lowest level of the prices of lemon oil (4 marks) ever attained, which was reached in March, was followed in May by a sharp movement upwards, bringing the quotations in a comparatively short time up to 5,25 marks per kilo.

This was brought about by very important purchases made here by order and on account of American firms. But the exaggerated demands of the proprietors, who would have been happy to avail themselves of this opportunity for recovering their previous losses, was followed by the usual reaction. When the first excitement was over, all purchasers who were not absolutely compelled to buy immediately, withdrew from the market, and very soon afterwards the prices began again to give way. At the same time, the movement had sufficed to remove the prices of this article for the current year definitely from the unsound level at which the manufacturers lost money, and to place them on a better and a more permanent basis. The above-mentioned weakening of the prices proceeded slowly, and when 4,50 marks had been reached they remained for a long time at that level, and only advanced again during the last few weeks when the available stocks of lemon oil had dwindled to about 25000 kilos; at the same time last year they were estimated at about 40000 kilos.

It seems safe to assume that these small stocks will have been used up by the time the new oil arrives, and that at the turn of the year there will be no stock worth mentioning.

With regard to the new crop, no exact estimates are as yet possible; on the whole it is expected that the result of the harvest will be about one-fifth below that of last year.

In a work entitled "La nuova adulterazione dell' essenza di limone"[1], Berté, with the help of very voluminous examination-material, discusses the adulteration of lemon oil, with special regard to the adulteration with lemon oil terpenes, which has increased very considerably during the last few years in consequence of the

[1] Estratto dal "Bolletino chimico-farmaceutico" No. 10, May **1904**.

increased production of concentrated oil of lemon. According to Berté, even slight adulterations of lemon oil with oil of turpentine or lemon oil terpenes can be detected by means of the method recommended by himself and Soldaini, by which $50^0/_0$ of the oil are distilled off and the rotations of the oil, the distillate, and the residue are compared with each other. This method is said to be decidedly more useful, especially in the case of oils adulterated with lemon oil terpenes, than the method recommended by us (test of the first $10^0/_0$ of the distillate) by means of which even an adulteration with $15^0/_0$ terpenes cannot be recognised.

Berté has, in the manner indicated above, distilled pure as well as purposely adulterated lemon oils, and arrived thereby at the following results:

	$a_{D\,20^o}$	$a_{D\,20^o}$ of the distillate $(50^0/_0)$	$a_{D\,20^o}$ of the residue
Pure lemon oil	$+62^o\,40'$	$+63^o\,45'$	$+61^o\,20'$
with $5^0/_0$ Americ. turpentine oil	$+59^o\,40'$	$+59^o\,30'$	$+59^o\,20'$
„ $5^0/_0$ French „ „ [1])	$+57^o\,45'$	$+57^o\,10'$	$+56^o\,30'$
„ $5^0/_0$ lemon oil terpenes .	$+63^o$	$+63^o\,10'$	$+62^o\,40'$
„ $10^0/_0$ „ „ „ . .	$+63^o\,20'$	$+63^o\,20'$	$+63^o\,25'$
lemon oil terpenes	$+66^o$	$+60^o\,10'$	$+70^o\,20'$

According to these results, the distillate of pure oil has a slightly higher rotatory power than the original oil, whilst the rotation of the residue is correspondingly lower. With adulterated oils the conditions are somewhat different. In the case of the grosser adulterations the differences naturally become more pronounced; small admixtures (as the results quoted show), cause but very slight differences, and it is questionable whether it is possible on the strength of the condition of the distillate, to arrive here in every case at a definite conclusion on the quality of the oil.

Contrary to Berté, we have even found no difference whatever between pure oil and oils adulterated with a small quantity of lemon oil terpene, when we repeated the tests according to Berté's method. We added to a lemon oil $10^0/_0$ lemon oil terpenes as obtained by us as a by-product in the manufacture of concentrated lemon oil, and tested the pure oil and the adulterated one in the manner indicated by Berté:

---

[1]) It is clear that in consequence of the distillation carried out at ordinary pressure, decompositions occur, as otherwise it would not be possible to explain that both the distillate and the residue have a lower rotatory power than the original oil.

	$a_{D\,20°}$	$a_{D\,29°}$ of the distillate	$a_{D\,20°}$ of the residue
Pure lemon oil	$+59°24'$	$+60°34'$	$+57°38'$
with 10°/₀ lemon oil terpenes	$+59°55'$	$+60°39'$	$+58°52'$
lemon oil terpenes	$+63°41'$	$+62°23'$	$+64°33'$

In this case Berté's method has failed, as the adulterated oil gives results which correspond with those of the pure oil.

We readily admit that the distillation method introduced by us in 1896, which of course is only directed against the adulteration with turpentine oil, at that time the only one practised, will equally fail to bring us to the end in view; but we only wish to show by the above example that the process recommended by Berté is not nearly as reliable as would appear from the tests carried out by himself.

When the opportunity arises we will return to this subject.

E. Schmidt[1]) has succeeded in clearing up the constitution of citroptene which has also been examined already by others. Citroptene, obtained from the crystalline distillation-residues of lemon oil by treatment with ether, is a granular crystalline mass insoluble in ether, which after repeated recrystallisation from acetone and methyl alcohol and also from dilute alcohol, with the addition of animal charcoal, was obtained in the form of brilliant colourless needles of the melting point 146° to 147°.

The solutions show a beautiful blue-violet fluorescence. The analysis gives the composition $C_{11}H_{10}O_4$, and the methoxyl-determination the content of two methoxyl-groups. When melted with potash, phloroglucinol and acetic acid were formed. With bromine in chloroform solution citroptene combines into a dibromide $C_{10}H_{10}Br_2O_4$ melting at 250° to 260°. The assumption that citroptene is a methylated dioxycoumarin, was confirmed by synthesis of the body. Starting from phloroglucinol, Schmidt, by conversion of this phenol into phloroglucinic aldehyde, and condensation of the latter according to the coumarin synthesis, obtained a dioxycoumarin which on methylation yielded a compound which corresponded in its properties with citroptene. Melting point 146° to 147°. The following formula therefore belongs to citroptene:

$$CH_3O \underset{H \quad O \cdot CO}{\overset{H \quad OCH_3}{\bigcirc}} CH = CH$$

Mandarin Oil. Owing to small stocks, this oil fetches good prices, whilst future oil, with bad prospects for the coming harvest, is offered at still higher quotations for forward delivery.

---

[1]) Arch. der Pharm. **242** (1904), 288.

Orange Oil, bitter, has been used up completely, and at last fetched quite enormous prices, up to 25 marks per kilo.

Orange Oil, sweet. The view to which we gave expression in our spring-Report, that the prices of this oil would soon advance again, has been fully confirmed; this was inevitable, as at that time already the available stocks were estimated at barely 7000 kilos.

A disquieting fact, which influenced the natural movement of the price, was that many local speculators selected this oil for the sphere of their activity; this was the cause that the brisk demand drove up the prices to 18 marks, and in some cases even to 19 marks; but it also led to a subsequent fall in the quotations to about 15,25 marks, in spite of the large shrinkage which had meanwhile taken place in the stocks.

It can be stated definitely that of last year's oil there is now nothing left over.

The estimates of the coming harvest are much lower than last year, both in Sicily and in Calabria; the crop of fruit is calculated at about $^1/_2$ or $^2/_8$ of that of the previous year. It should also be borne in mind that the traffic of loose fruit in truck-loads by rail to Upper Italy is constantly increasing, and it is already reported that the Upper Italian firms have made large contracts for forward delivery in Calabria and also in Sicily. The fruit used for this purpose is, however, just the kind which does not keep so well, and is neither entirely faultless nor suitable for shipment by sea; for these reasons such fruit had up to now been used for the production of oil.

Under these conditions the expectation in the producing country is, that considerably higher prices will rule than was the case at the same time last year.

Eucalyptus Oil. The lion's share of the trade again belongs to Australia, since the Algerian distillers have been found to be unreliable. Victoria exported in 1902 eucalyptus oil of a total value of £ 14,127, equal to at least 250000 kilos. Portugal also supplies a few thousand kilos of a good normal distillate.

Australian Eucalyptus Oils. To the great kindness of Messrs. Baker and Smith, Curators of the *Technological Museum* of *Sydney,* N. S. W., we are indebted for a collection of Australian eucalyptus oils which in their richness testify in a remarkable degree to the zeal and success with which the study of the so extremely varied family of the eucalypts is carried on in Australia.

This collection is all the more valuable, as details are given with each oil on the mother-plant and the yield of oil, and also on the principal chemical constituents. Owing to the interest which these data may justly claim, we reprint them on pp. 32 to 37, and for the sake of greater clearness we have arranged them in tabular form.

No.	Botanical name	Vernacular name	Origin	Average yield of oil in %	Principal chemical constituents
1.	*E. tessellaris* F. v. M.	Moreton Bay, Ash	Narrabri, N. S. W.	0,151	Pinene, sesquiterpene
2.	*E. trachyphloia* F. v. M.	Bloodwood	Murrumbo, N. S. W.	0,199	Pinene, sesquiterpene
6 [1])	*E. eximia* R. T. B.	White Blood-wood	Springwood, N. S. W.	0,462	Pinene
7.	*E. botryoides* Sm.	Bastard Mahogany	Milton, N. S. W.	0,086	d-Pinene
8.	*E. robusta* Sm.	Swamp Mahogany	La Perouse, N. S. W.	0,161	Pinene
9.	*E. saligna* Sm.	Blue Gum	Gosford, N. S. W.	0,241	Pinene
10.	*E. nova-anglica* D. & M.	Black Peppermint	Walcha, N. S. W.	0,51	Terpenes
11.	*E. umbra* R. T. B.	A Stringy-bark	Lismore, N. S. W.	0,1615	Pinene, also an acetic acid ester
12.	*E. dextropinea* R. T. B.	Stringybark	Barber's Creek, N. S. W.	0,798	d-Pinene
13.	*E. Wilkinsoniana* R. T. B.	Small Leaved Stringybark	Barber's Creek, N. S. W.	1,01	l-Pinene
14.	*E. laevopinea* R. T. B.	Silver Top Stringybark	Rylstone, N. S. W.	0,66	l-Pinene
15.	*E. Bäuerleni* F. v. M.	Brown Gum	Monga, N. S. W.	0,328	Terpenes, eucalyptol, chiefly the former
16.	*E. propinqua* D. & M.	Grey Gum	Woodburn, N. S. W.	0,235	
17.	*E. affinis* D. & M.	Black Box	Grenfell, N. S. W.	0,259	
18.	*E. paludosa* R. T. B.	Swamp Gum	Barber's Creek, N. S. W.	0,197	
19.	*E. lactea* R. T. B.	Spotted Gum	Ilford, N. S. W.	0,557	Terpenes
21.	*E. intertexta* R. T. B.	Gum or Red Gum	Nyngan, N. S. W.	0,395	Pinene, eucalyptol

[1]) The missing numbers are also wanting in the original.

No.	Botanical name	Vernacular name	Origin	Average yield of oil in %	Principal chemical constituents
22.	*E. maculata* Hook.	Spotted Gum	Currawang Creek, N. S. W.	0,169	Pinene, eucalyptol, chiefly the former
25.	*E. quadrangulata* D. & M.	Grey Box	Milton, N. S. W.	0,684	Pinene, eucalyptol
26.	*E. conica* D. & M.	Box	Parkes, N. S. W.	0,587	Pinene, eucalyptol
27.	*E. Bosistoana* F. v. M.	Box	Barber's Creek, N. S. W.	0,968	Pinene, eucalyptol, chiefly the former
28.	*E. eugenioides* Sieb	White Stringybark	Canterbury, N. S. W.	0,742	Terpenes, eucalyptol
30.	*E. resinifera* Sm.	Mahogany	Gosford, N. S. W.	0,302	
31.	*E. polyanthema* Sieb	Red Box	Pambula, N. S. W.	0,825	Pinene, eucalyptol
32.	*E. Behriana* F. v. M.	Mallee Gum	Wyalong, N. S. W.	0,614	
33.	*E. Rossi* R. T. B. and H. G. S.	White or Brittle Gum	Bathurst, N. S. W.	0,65	Pinene, eucalyptol, also piperitone[1]
34.	*E. pendula* A. Cunn.	Red Box	Nyngan, N. S. W.	0,67	Pinene, eucalyptol, chiefly the latter
35.	*E. dealbata* A. Cunn.	Cabbage or Montain Gum	Condobolin, N. S. W.	0,856	Pinene, eucalyptol, chiefly the latter
37.	*E. rostrata* Schl. *var. borealis* R. T. B. and H. G. S.	River Red Gum	Nyngan, N. S. W.	1,001	Pinene, eucalyptol
38.	*E. maculosa* R. T. B.	Spotted Gum	Bungendore, N. S. W.	0,846	Pinene, eucalyptol
39.	*E. camphora* R. T. B.	Sallow	Delegate, N. S. W.	0,836	Eucalyptol
40.	*E. punctata* D. C.	Grey Gum	Canterbury, N. S. W.	0,781	Pinene, eucalyptol

[1]) With the name "piperitone" Baker and Smith designate a constituent with a peppermint-like odour, which is present in various eucalyptus oils. For further details, see Baker and Smith: "A research on the eucalypts, especially in regard to their essential oils", Sydney 1901, page 229.

No.	Botanical name	Vernacular name	Origin	Average yield of oil in %	Principal chemical constituents
41.	*E. squamosa* D. & M.	Ironwood	National Park, N. S. W.	0,643	
42.	*E. Bridgesiana* R. T. B.	Apple or Woollybutt	Ilford, N. S. W.	0,619	Eucalyptol
43.	*E. goniocalyx* F. v. M.	Montain Gum	Monga, N. S. W.	0,881	
44.	*E. bicolor* A. Cunn.	Bastard Box	St. Mary's, N. S. W.	0,52	
45.	*E. viminalis* Var. (a)		Crookwell, N. S. W.	0,701	Pinene, eucalyptol, benzaldehyde (?)
46.	*E. populifolia* F. v. M.	Poplar Leaved Box	Nyngan, N. S. W.	0,758	
47.	*E. longifolia* Link.	Woollybutt	Canterbury, N. S. W.	0,535	
48.	*E. Maideni* F. v. M.	Blue Gum	Barber's Creek, N. S. W.	1,304	Eucalyptol
49.	*E. globulus* Labill.	Blue Gum	Jenolan, N. S. W.	0,745	
50.	*E. pulverulenta* Sims.		Bathurst, N. S. W.	2,22	
51.	*E. cinerea* F. v. M.	Argyle Apple	Barber's Creek, N. S. W.	1,20	Eucalyptol, some valeric acid ester
51a.	*E. cordata* Labill.		Tasmania	2,32	
54.	*E. Morrisii* R. T. B.	Grey Mallee	Gerilambone, N. S. W.	1,69	Eucalyptol
55.	*E. Smithii* R. T. B.	Gully Ash or White Top	Monga, N. S. W.	1,434	
56.	*E. sideroxylon* A. Cunn.	Red Flowering Ironbark	Liverpool, N. S. W.	0,537	Pinene, eucalyptol
57.	*E. cambagei* D. & M.	Bastard Box or Bundy	Bathurst, N. S. W.	0,735	Eucalyptol, some aromadendral [1]
58.	*E. polybractea* R. T. B.	Blue Mallee	Wyalong, N. S. W.	0,135	Pinene, eucalyptol, aromadendral

[1] With regard to aromadendral, comp. Report April **1901**, 33, October **1901**, 28, and October **1903**, 35.

No.	Botanical name	Vernacular name	Origin	Average yield of oil in %	Principal chemical constituents
59.	E. dumosa A. Cunn.	White Mallee	Coolabah, N. S. W.	1,00	Terpenes, eucalyptol, aromadendral
60.	E. oleosa F. v. M.	Red or Water Mallee	Nyngan, N. S. W.	0,97	Pinene, eucalyptol, aromadendral
61.	E. cneorifolia D. C.		Kangaroo Island		Pinene, eucalyptol, aromadendral
62.	E. stricta Sieb.	Mountain Mallee	Blue Mountains, N. S. W.	0,494	Eucalyptol
63.	E. melliodora A. Cunn.	Yellow Box	Rylstone, N. S. W.	0,676	Pinene, eucalyptol, frequently also phellandrene
64.	E. ovalifolia, var. lanceolata R. T. B.	Red Box	Camboon, N. S. W.	0,579	Pinene, eucalyptol, phellandrene
64a.	E. Risdoni Hook. f.	Risdon or Drooping Gum	Tasmania	1,348	Eucalyptol, phellandrene, piperitone
66.	E. punctata D. C., var. didyma R. T. B. & H. G. S.		Barber's Creek, N. S. W.	0,402	Pinene, eucalyptol, aromadendral
67.	E. gracilis F. v. M.	A Mallee	Gunbar, N. S. W.	0,901	Pinene, eucalyptol, aromadendral; the terpenes predominate
68.	E. viridis R. T. B.	Green Mallee	Gerilambone, N. S. W.	1,06	Terpenes, a small quantity aromadendral
69.	E. Woollsiana R. T. B.	Mallee Box	Gerilambone and Nyngan, N. S. W.	0,449	Aromadendral
70.	E. albens Mig.	White Box	Rylstone, N. S. W.	0,101	Aromadendral
71.	E. hemiphloia F. v. M.	Box	Belmore, N. S. W.	0,554	Pinene, eucalyptol, aromadendral
72.	E. viminalis Labill.	Manna Gum	Cadia, N. S. W.	0,354	Phellandrene, eucalyptol
73.	E. rostrata Sch.	Murray Red Gum	Albury, N. S. W.	0,299	Chiefly terpenes, also eucalyptol and aromadendral; sometimes phellandrene
74.	E. ovalifolia R. T. B.		Rylstone, N. S. W.	0,216	Pinene, eucalyptol, phellandrene

No.	Botanical name	Vernacular name	Origin	Average yield of oil in %	Principal chemical constituents
75.	E. Dawsoni R. T. B.	Slaty Gum	Bylong, N. S. W.	0,172	Phellandrene, sesquiterpene
76.	E. angophoroides R. T. B.	Apple Topped Box	Towrang, N. S. W.	0,185	Terpenes, among which phellandrene
77.	E. fastigata D. & M.	Cut Tail	Monga, N. S. W.	0,263	Pinene, phellandrene, eudesmol
78.	E. macrorhyncha F. v. M.	Red Stringy-bark	Rylstone, N. S. W.	0,272	Terpenes, eucalyptol, eudesmol
79.	E. capitellata Sm.	Brown Stringybark	Canterbury, N. S. W.	0,103	Terpenes, small quantities of eucalyptol
80.	E. nigra R. T. B.	Black Stringybark	Woodburn, N. S. W.	0,0295	Phellandrene
81.	E. pilularis Sm.	Blackbutt	Belmore, N. S. W.	0,13	Terpenes, also an as yet unknown alcohol
83.	E. acmenoides Sch.	White Mahogany	Lismore, N. S. W.	0,358	Terpenes, among which phellandrene
84.	E. fraxinoides H. D. & J. H. M.	White Ash	Monga, N. S. W.	0,985	Terpenes, chiefly phellandrene
85.	E. Fletcheri R. T. B.	Lignum Vitae or Box	Thirlmere, N. S. W.	0,352	Terpenes, chiefly phellandrene
86.	E. microtheca F. v. M.	Coolybah or Tangoon	Narrabri, N. S. W.	0,150	
87.	E. haemastoma Sm.	White or Scribbly Gum	Barber's Creek, N. S. W.	0,241	Phellandrene, sesquiterpene
89.	E. crebra F. v. M.	Narrow Leaved Iron-bark	Rylstone, N. S. W.	0,159	Pinene, phellandrene, eucalyptol
92.	E. piperita Sm.	The Sydney Peppermint	Gosford, N. S. W.	0,627	Pinene, phellandrene, eucalyptol, eudesmol and piperitone
93.	E. amygdalina Labill.	Messmate	Moss Vale, N. S. W.	3,393	Phellandrene, eucalyptol and piperitone
94.	E. vitrea R. T. B.	White Top Messmate	Crookwell, N. S. W.	1,48	Phellandrene, eucalyptol
95.	E. Luehmanniana F. v. M.		National Park, N. S. W.	0,289	Phellandrene
96.	E. coriacea A. Cunn.	Cabbage Gum	Ilford, N. S. W.	0,452	Phellandrene, piperitone

No.	Botanical name	Vernacular name	Origin	Average yield of oil in %	Principal chemical constituents
97.	E. Sieberiana F. v. M.	Mountain Ash	Barber's Creek, N. S. W.	0,421	
98.	E. oreades R. T. B.	A Mountain Ash	Lawson, N. S. W.	1,16	
99.	E. dives Sch.	Broad Leaved Peppermint	Fagan's Creek, N. S. W.	2,233	Phellandrene, piperitone
100.	E. radiata Sieb.	White Top Peppermint	Monga, N. S. W.	1,641	
101.	E. delegatensis R. T. B.	White Ash, Silver Top Mountain Ash	Delegate Mountain, N. S. W.	1,76	
102.	E. obliqua L'Hér.	Stringybark	Monga, N. S. W.	0,677	Phellandrene, a small quantity aromadendral
103.	E. stellulata Sieb.	Lead Gum	Rylstone, N. S. W.	0,293	Phellandrene
104.	E. Macarthuri H. D. and J. H. M.	Paddy River Box	Wingello, N. S. W.	0,112	Geranyl acetate
106.	E. virgata Sieb.		Springwood, N. S. W.	0,283	Eudesmol
107.	E. patentinervis R. T. B.	Mahogany	Belmore, N. S. W.	0,254	Terpenes, citral and an as yet unknown alcohol
108.	E. apiculata R. T. B. and H. G. S.		Berrima, N. S. W.	0,296	Terpenes, piperitone
109.	E. citriodora Hook.	Citron Scented Gum	Sydney, N. S. W.	0,586	Citronellal

On the therapeutic value of eucalyptus oils, with special regard to their antiseptic properties, Cuthbert Hall[1]) has published a work carried out by him in the Pathological Institute of Sydney University.

The tests naturally were made not only with the oils, but above all also with their more important constituents known up to the present, in order to determine the proper value of each.

---

[1]) On Eucalyptus oils especially in relation to their bactericidal power. Parramatta, N. S. W. 1904.

For the bacteriological experiments two different bacteria of high resisting power were employed, viz., the stronger *Staphylococcus pyogenes aureus,* and the weaker *Bacillus coli communis.*

The experiments made by Hall showed that eucalyptol as such, with regard to its antiseptic properties, is inferior to all other constituents of the eucalyptus oils, and that aromadendral, piperitone[1]), and phellandrene are here particularly active.

For example, *Bacillus coli communis* is only destroyed after 8 hours by eucalyptol, whilst the same effect is obtained with aromadendral in 10 minutes, with piperitone in 40 minutes, and with phellandrene in $1^1/_2$ hours. Somewhat less powerful, but stronger than eucalyptol, are d- and l-pinene and aromadendrene. Eudesmol also proved to be a strongly antiseptic substance, as it enhanced the activity of eucalyptol and piperitone, which served as solvents for this body.

The conditions are decidedly more favourable for eucalyptol when it contains ozone which is formed by the slow oxidation of the terpenes, especially phellandrene and aromadendrene[2]). Under such conditions the vitality of *Bacillus coli communis* ceases already after 15 minutes, so that here eucalyptol is only exceeded in activity by aromadendral.

The results of the action on *Staphylococcus* were entirely analogous.

The results of the tests with eucalyptus oils naturally correspond to the content of the above-named constituents in the oils. To this, however, a further essential factor must be added, consisting of the presence of free acid (acetic acid) and consequently of a possible content of iron or copper originating from the vessels in which the oil is kept, by which the antiseptic activity of the oils is increased in a marked degree. However, as a rule this does not carry much weight, as the oils intended for medicinal purposes are always rectified and neutral.

As it is for the rest still an open question whether oil containing aromadendral can be used with advantage for medicinal purposes, and as moreover eucalyptol is the constituent which occurs chiefly in eucalyptus oil, Hall arrives at the conclusion that in the first place "ozone" is of very considerable importance for the antibacterial power of eucalyptus oils, and that the oils, in order to enhance their antiseptic properties, should be "ozonised" by allowing light and air to act upon them for a prolonged period (at least two months); this is done by closing only loosely with plugs of cotton-wool the vessels in which the oil is kept, and frequently shaking up the oils which are exposed to sunlight.

---

[1]) With regard to piperitone, see p. 33, note.

[2]) It is a long-known fact that in the slow oxidation of terpenes not ozone, but organic peroxides and hydrogen peroxide occur. If further on we continue to speak of "ozonised" eucalyptol, etc: this is only done for the sake of simplicity. Comp. Gildemeister and Hoffmann, "The Volatile Oils", p. 231.

For the same reason only ozonised oils should be used for the production of eucalyptol, in order to obtain in this manner also an active eucalyptol. The ozone once present in the eucalyptol, is not abstracted from the latter (?) by the method of manufacture (freezing out or rectification).

Finally Hall discusses briefly the pharmacology of eucalyptus oils, where he again lays stress on the importance of the ozone-content.

Eucalyptus oil is frequently used with advantage both externally and internally. It is said to render good services as an inhalation remedy in diphtheria, scarlet-fever, whooping-cough, bronchial catarrh, pneumonia, and influenza. In the form of an embrocation it is useful in rheumatism, and it is also employed in gonorrhœa and leucorrhœa. But the action against cancer which has been ascribed to it, has not been confirmed.

The internal administration of eucalyptus oil takes place either per os, or hypodermically; in the latter case, olive oil and glycerin are used as diluents. When employed hypodermically, it has had an absolutely specific action in several cases of pyæmia, puerperal fever, and septicæmia; equally favourable results have been observed in erysipelas. In every case it simultaneously lowers the temperature.

Administered per os it is said to be successful in bronchitis, phthisis, and scarlet fever; in typhus Hall has also tried the oil, and, according to the results obtained up to the present, it appears to be useful also in such cases. That the oil or leaves of eucalyptus are largely used against malaria is well known. We would still mention that an infusion of the leaves of *Eucalyptus globulus* is said to be very active in diabetes, whilst the corresponding oil produces no effect.

Eucalyptol is probably free from toxic action, as in doses of 10 gm. daily of an oil rich in eucalyptol, no disturbances worth mentioning have been observed.

As a continuation of the excellent monograph of the eucalyptus published by J. H. Maiden, of which we discussed the first 3 parts in our Report of October 1903 (p. 38), the 4th part has now made its appearance[1]). In this part, *Eucalyptus incrassata* Labillardière and *E. foecunda* Schauer are described. As a typical form of the first-named, *E. dumosa* A. Cunn. var. *scyphoralyx* F. v. M. is indicated. The following varieties are also mentioned: var. *dumosa* F. v. M. (synonyms: *E. dumosa* A. Cunn., possibly with var. *punctilulata* Benth. and var. *rhodophloia* Benth., *E. lamprocarpa* F. v. M., *E. Muelleri* Miq., *E. glomerata* Tausch). Var. *conglobata* R. Br. (Synonyma: *E. conglobata* R. Br., *E. anceps* R. Br., *E. pachyphylla* F. v. M.). Var. *angulosa* Benth.

---

[1]) A critical revision of the genus Eucalyptus, Part IV, Sydney, 1904.

(synonyms: *E. angulosa* Schauer, *E. cuspidata* Turcz., *E. costata* R. Br.,
E. linopoda R. Br., *E. rugosa* R. Br., *E. sulcata* Tausch, *E. pachyphylla*
A. Cunn.). Var. *goniantha* var. nov. (*E. goniantha* Turcz.). Var. *grossa*,
var. nov. (synonyms: *E. grossa* F. v. M., *E. pachypoda* F. v. M.). As
synonyms of *E. foecunda* Schauer, Maiden mentions: Var. *loxophleba*
J. G. Luehmann with *E. loxophleba* Benth. and *E. amygdalina* Schauer
non Labill. A large number of well-executed illustrations is added
to this Part.

Extract Oils, essential. It is well known that in the blossom
industry of the South of France the odour of many blossoms which
are important for purposes of perfumery, has now for some years
been obtained on an extensive scale by extraction with light petroleum. The products of this manufacturing process, the blossom-ex-
tracts, are distinguished by their natural aroma, but in addition to the
essential oils which alone come under consideration for the odour,
they contain large quantities of odourless constituents such as vegetable
waxes, resins, paraffins, etc., of which many dissolve only with difficulty
in alcohol. This odourless ballast can partly be removed from the
extracts by treatment with alcohol, or more completely by distillation
with water-vapour. In view of the comparatively small quantity of
essential oil contained in the extracts, a large proportion of the oil
passes over in the distillation-water, from which it is recovered by
extraction with salt and with ether. The pure essential oils of the
blossoms which remain behind after distilling off the ether, contain
the odorous substances of the blossoms in a concentrated form, and
they have for several years occupied in a marked degree the attention
of the chemists whose work lies in the domain of the essential oils.
H. von Soden[1]) describes a number of blossom-oils obtained by this
process, some of which are already known, and of which others have
not yet been examined. 1000 kilos violet-blossoms (March 1903)
yielded 31 gm. (0,0031%) of a non-fluorescent essential oil with a
faint greenish colour, which did not solidify in a freezing mixture;
$d_{15°}$ 0,920; $a_{D17°} + 104°$ 15'; acid number 10; ester number 37; readily
soluble in alcohol. In a concentrated form essential oil of violets has
only a feeble violet-odour; only when diluted 1 : 5000 to 10000 this
odour becomes prominent, and is then accompanied by a herb-like
additional odour originating from the sepals of the blossom. Oil of
violets is the most expensive oil used in practice. The production of
the oil freed from the odourless admixtures would cost £ 4,000 per
kilo, without reckoning manufacturing charges.

From 600 kilos mignonette-blossoms (June 1903) 0,003% essential
oil of mignonette-blossoms was obtained. The yellow oil, possessing

---

[1]) Journ. f. prakt. Chemie II. **69** (1904), 256.

an intense mignonette odour, solidifies in the cold[1]), is not fluorescent, and shows the following constants: d_{15^o} 0,961; $a_{D17^o} + 31^o 20'$; acid number 16; ester number 85.

With alcoholic potash the oil develops bases of an ammonia-like odour; moreover the presence of aldehydes was observed. To produce 1 kilo oil, 33000 kilos blossoms, value ℒ 1,500, would be required.

1300 kilos orange-blossoms (April 1902) yielded 0,06% essential oil of the following constants: d_{15^o} 0,9245; $a_D - 2^o 30'$; acid number 4; ester number 102; content of anthranilic acid ester 6,90 % [2]).

8000 kilos French rose-blossoms yielded 0,052 % essential oil of the solidifying point 5° to 7°; d_{15^o} 0,967; $a_{D17^o} - 1^o 55'$; ester number 4,6; ester number after acetylation 295. By means of phthalic acid anhydride, 75 to 80% alcohols were isolated from the oil, of which 75% consisted of phenyl ethyl alcohol, and 25% of the other primary alcohols present in rose oil.

45 kilos German roses yielded 0,0107 % essential extract-oil. Solidifying point + 12°; d_{19^o} 0,984; $a_{D17^o} + 0^o 9'$; acid number 3; ester number 4; ester number after acetylation 313,5; content of phenyl ethyl alcohol 75%, of primary aliphatic alcohols 15 % [3]).

2000 kilos jasmine-blossoms (July, August) yielded 0,077% jasmine-blossom oil; d_{15^o} 0,9955; $a_D - 1^o$; acid number 2,5; ester number 190 = 51 % benzyl acetate. In September/October, there was obtained from 1800 kilos blossoms 0,0718 % oil which had slightly different constants; d_{15^o} 0,967; a_D faintly to the left; acid number 3,5; ester number 161 = 43,3 % benzyl acetate. The oil contained a fairly large quantity indol and, judging by the fluorescence, also methyl ester of anthranilic acid. Contrary to this is Hesse's observation, who had obtained jasmine-blossom oils free from indol both by distillation of jasmine-blossoms, and by extraction. The calculation of the yields also shows a striking difference, as the quantity of essential oil obtained by v. Soden exceeds by nearly four times that ascertained by Hesse and by Pillet[4]).

From 1000 kilos cassie-blossoms (*Acacia farnesiana*, October to December), 840 grams = 0,084% essential oil were obtained;

---

[1]) We have previously obtained by distillation of fresh mignonette-blossoms with water vapour, 0,002 % essential oil of mignonette-blossoms, of the consistency of orris oil. Report October **1891**, 40.

[2]) Comp. Hesse and Zeitschel, Journ. f. prakt. Chemie II. **66** (1902), 512; Report October **1903**, 50.

[3]) The essential extract oil from German rose-blossoms was produced by us already in 1900, and a content of about 62 % phenyl ethyl alcohol ascertained in it. Berl. Berichte **33** (1900), 2302; Report October **1900**, 54.

[4]) Comp. A. Hesse, Berl. Berichte **37** (1904), 1457, also the present Report p. 49.

solidifying point $18°$ to $19°$; $d_{27°}$ 1,040; $a_{D25°}$ — $0°40'$; acid number 42,50; ester number 114 $= 30,9^0/_0$ methyl ester of salicylic acid[1]).

Fennel Oil. A question which was submitted to us some considerable time ago, as to the cause of the occurrence that fennel oil is sometimes heavier than water, induces us to return to this subject in this place. When fennel oil[2]) is kept in an unsuitable manner, that is to say when light and air are admitted, a gradual increase in the specific gravity is observed which may finally bring the latter above 1.

This increase in the weight is due to oxidation-processes, or occurrences of polymerisation, as the anethol present in the fennel oil in the presence of light and air partly oxidises into anisic aldehyde and anisic acid, and partly becomes polymerised. The products in question now not only cause an increase in the specific gravity, but also at the same time a greater solubility of the oil. Moreover, the solidifying point is naturally greatly influenced, and finally such oils no longer solidify at all.

Whereas normal fennel oil has the following properties: $d_{15°}$ 0,965 to 0,975; solidifying point $+3°$ to $+6°$, sometimes over $+8°$; soluble in 6 to 8 vol. and more 80 per cent. alcohol, possibly with cloudiness, we found for an oil which had changed in the manner described above, $d_{15°}$ 1,0053; solidifying point $—8°$; soluble in 3 vol. and more 80 per cent. alcohol.

These differences become still more pronounced in the case of pure anethol. An anethol which had been exposed for two years to the influence of light and air, had undergone a remarkable change, as will be seen from the following comparison:

I. Original anethol: $d_{25°}$ 0,9846; $a_D \pm 0°$; $n_{D25°}$ 1,56079; solidifying point $+21°3'$; soluble in 2 and more vol. 90 per cent. alcohol.

II. The same anethol after having been kept for two years, light and air being admitted: $d_{25°}$ 1,1245: $a_D \pm 0°$; $n_{D25°}$ 1,54906; does not yet solidify at $—20°$; soluble in 1,5 to 2 and more vol. 70 per cent. alcohol.

The changed oil had a yellow colour and was less mobile than normal anethol; the taste was no longer like anise, but disagreeable and bitter. By treatment with bisulphite, anisic aldehyde could be isolated, but only traces of anisic acid could be detected. The bulk of the anethol had become polymerised.

---

[1]) Comp. our work on the essential oil of acacia blossoms, Report October **1903**, 18. Journ. f. prakt. Chem. II. **68** (1903), 749.

[2]) The same applies also to anise oil. Comp. Gildemeister and Hoffmann, "The Volatile Oils", page 561.

The foregoing shows once more how important it is especially in the case of essential oils, that they should be kept in a suitable manner. It is absolutely necessary, particularly if the oils are kept for a prolonged time, to protect them as much as possible from exposure to light and air.

Geranium Oil. During the last few years, the cultivation of geraniums has been increasing considerably in Algeria, the result of which is noticeable in the large quantities of the distillate now offered. The prices have gone down about 5 fr. per kilo in the course of the last six months. The opinion expressed in our last Report on the present year's qualities has been confirmed; the new oils are of exceptionally good quality.

The geranium plantations in Réunion were totally destroyed on the 21st and 22nd March of this year, by a cyclone which has not been equalled in violence since 1863, and there was every reason to believe that the prices of geranium oil would advance considerably. But, contrary to all calculations, the opposite has taken place; after a rise which continued only for a few weeks, the quotations are now lower than before the catastrophe, and that owing to the fact that the new geranium plantations already yield crops in four months' time, whilst all other cultivated plants require a longer period. For this reason the geranium cultivations were restored before any others, and that on a large scale, and they have now for several months again been capable of producing crops.

Moreover, large stocks of oil had accumulated in Réunion and in France, and even the S. S. "Iraouaddy", which sailed on the 26th March, landed at Marseilles no less than 141 cases containing about 1800 kilos geranium oil. In the month of August, the following quantities of oil, the bulk of which may possibly originate from the new plantations, arrived at Marseilles: —

per S.S. "Melbourne"	35	cases
„ „ "Natal"	112	„
„ „ "Madagascar"	5	„
Total	152 cases =	1800 kilos.

The present stock at Marseilles is estimated at about 4000 kilos.

After what has been said, it is hardly to be expected that a rise in the prices will take place in the immediate future, but a further fall seems also out of the question. Our sales of this kind of geranium oil are exceptionally large, and amount to a considerable proportion of the whole annual production.

The fine East Indian geranium or Palmarosa oil was during the last few months quoted even a little cheaper than at the beginning of the season, and has rightly been bought largely on speculation. The present value of this important oil is exceptionally low, a natural consequence of a greatly increased production due to the previous high prices. At the present quotations, however, the distillation has become unremunerative, and it is not improbable that the prices will advance to a more equitable level at an early date.

It is now an exceptionally favourable opportunity for the purchase of fine Gingergrass oil, of a quality such as has not been met with in commerce for years. Our examinations have proved that it is a pure distillate deserving absolute confidence.

On the cultivation of *Pelargonium capitatum* Ait., Jacob de Cordemoy[1]) reports in an interesting work from which we abstract the following:

The cultivation of geranium, which has considerably increased of late years, is carried on at Réunion at altitudes of 1300 to 4000 feet; in the regions situated at a greater elevation the winter-cold is too severe, so that the plants are destroyed by frost. The most suitable soil for the cultivation of geranium is one rich in humus, such as is found in the virgin soil at the above-mentioned altitudes. For the production of the oil the distillers employ stills of simple construction, of which some 250 are in operation in the whole colony. Whilst up to last year this industry was not taxed and only an official permit was required for the erection of the stills, there has lately been imposed an annual tax of 10 fr. per still. About 700 to 1000 kilos leaves yield 1 kilo oil. The leaves after distillation are usefully employed in the form of manure.

Gingergrass Oil. A short time ago[2]) we reported on a new alcohol $C_{10}H_{16}O$, which forms an important constituent of gingergrass oil. From what we then said it was clear that the production of this body in the pure state, in consequence of the admixture of geraniol, is difficult, and that we had up to then only partly succeeded in carrying it out. We believe that we have now obtained a purer product in this manner, that the alcohol isolated from the saponified oil by fractional distillation was heated on a water-bath with double the quantity of 99 per cent. formic acid just to $75°$ to $80°$, and the formate thereby produced worked up in the well known manner for the alcohol. Whether a quantitative removal of the geraniol was thus accomplished is a matter which could not be decided with absolute certainty. But from the combustion and also from the physical constants it may be concluded that the resulting product was purer than that previously obtained.

---

[1]) Revue des cultures coloniales **14**' (1904), 170.
[2]) Report April **1904**, 55.

Boiling point $94,5°$ to $96°$ (4 to 5 mm. pressure), $228°$ to $229°$ (755 mm. pressure); $d_{15°}$ $0,9536$; $a_D + 12° 5'$; $n_{D20°}$ $1,49761$.

Combustion: $0,1424$ gm. of the substance: $0,4108$ gm. CO_2, $0,1352$ gm. H_2O.

	Found	Calculated for $C_{10}H_{16}O$
C	78,68 %	78,95 %
H	10,55 %	10,52 %

The analysis shows that the alcohol has the formula $C_{10}H_{16}O$. A crystalline derivative has not yet been obtained. For the purpose of ascertaining its constitution, it was attempted to produce oxidation and reduction products. The oxidation with 1 per cent. permanganate solution progressed so violently that it was necessary to cool with ice. The final product formed a thick syrup which could not even *in vacuo* be distilled without decomposition. For this reason we abandoned for the time the further examination. By means of Beckmann's mixture the alcohol was oxidised at the temperature of the water-bath, on the one hand into an aldehyde $C_{10}H_{14}O$, on the other into an acid of the melting point $132°$. The first-named product could be purified with the help of its bisulphite compound, and thus formed an oil with a cumin-like odour, of the boiling point $235°$. Its semicarbazone of the melting point $198°$ to $198,5°$, crystallising from alcohol in the form of leaflets, served us for analysis:

$0,1733$ gm. of the substance: $0,4028$ gm. CO_2, $0,1272$ gm. H_2O.

	Found	Calculated for $C_{11}H_{17}N_3O$
C	63,39 %	63,77 %
H	8,15 %	8,21 %.

The quantity of the acid was not sufficient for analysis. With the view of reducing the alcohol to a saturated hydrocarbon, we heated 5 gms. each with 20 gms. hydriodic acid (spec. gr. 1,7) and 3 gms. phosphorus in a sealed tube for 4 hours to $200°$. In addition to considerable quantities of resin, this resulted in a hydrocarbon readily volatile with water-vapour, which had an odour like cymene. Its boiling point lay at $172°$ to $179°$. Oxidation with concentrated permanganate solution at the temperature of the water-bath, led, in addition to terephthalic acid, also to p-oxyisopropylbenzoic acid of the melting point $155°$ to $156°$, that is to say, to compounds which are formed on oxidation of p-cymene. In spite of all this we do not at present wish to decide the question whether we had here to deal actually with p-cymene, or with a hydrocarbon yielding the same oxidation products. The occurrence of the same hydrocarbon was also observed when we treated the alcohol for some time at $100°$ with strong formic acid. For its isolation, the readily volatile constituents were driven over from the reaction mixture with steam. The distillate was then oxidised with dilute permanganate solution in the cold, until no further discoloration could be observed when more chameleon solution was added. Boiling point and oxidation products of the hydrocarbon corresponded with those given above. According to this result it is not very probable that in the foregoing case a reduction has taken place. It would rather seem that the formation of the hydrocarbon must be attributed in both cases to the water-abstracting action of the formic and hydriodic acids at increased temperature.

The further examination of gingergrass oil has still resulted in the following:

. d-limonene and dipentene. These two terpenes were both detected in the portions boiling between $175°$ and $180°$. A fraction of the boiling point $176°$ to $178°$ ($d_{15°}$ 0,8499, $\alpha_D + 42° 40'$), distilled over sodium, formed a nitrosochloride of the melting point $103°$ to $104°$, which was converted on the one hand with benzylamine into a-dipentene-nitrol benzylamine of the melting point $110°$, on the other, with piperidine, into a-dipentene-nitrol piperidide of the melting point $153°$. The production of the tetrabromide of the melting point $125°$ also proved the presence of dipentene. We were also able, on the repeated formation of the nitrosochloride from a fraction of the melting point $178°$ to $180°$ ($\alpha_D + 43° 52'$), to identify limonene with the help of its a-piperidine and a-benzylamine derivatives of the same melting point $93°$.

Aldehyde $C_{10}H_{16}O$. The oil portions passing over above the boiling point of the terpenes were found to contain small quantities of an aldehyde, when fuchsin sulphurous acid or sodium bisulphite liquor were added. In order to obtain this aldehyde, the fractions boiling between $80°$ and $90°$ (6 mm. pressure) were shaken for a prolonged time with bisulphite liquor, and the double compound separating off as a slimy mass was removed from the oil by absorption. From this bisulphite compound purified by repeated washing with alcohol and ether, the aldehyde was liberated by heating with soda solution, driven off with steam, and fractionated *in vacuo*. It represents a colourless oil of the following properties:

boiling point $76°$ to $78°$ (5 mm. pressure), $221°$ to $224°$ (754 mm. pressure); $d_{15°}$ 0,9351; $\alpha_D \pm 0°$; $n_{D20°}$ 1,47348.

It has a peculiar, not unpleasant odour, which reminds both of œnanthic aldehyde and of citronellal. We produced the following derivatives: the semicarbazone of the melting point $169°$ to $170°$; the semioxamazone of the melting point $244°$ to $245°$; the phenylhydrazone of the melting point $63°$, the β-naphthocinchoninic acid of the melting point $261°$, and the oxime of the melting point $115°$ to $116°$. The analysis of the semicarbazone shows that the formula $C_{10}H_{16}O$ belongs to the aldehyde.

0,1687 gm. of the substance: 0,3906 gm. CO_2, 0,1383 gm. H_2O.

	Found:	Calculated for $C_{11}H_{19}N_3O$:
C	63,15 %	63,16 %
H	9,11 %	9,09 %

The content of this aldehyde (isomeric with citral) may possibly not exceed 0,2 % of the oil. The quantity which we had in hand was obtained from a large quantity of oil, and amounted to about 95 gm. The aldehyde can be reduced with comparative ease to the corresponding alcohol, by means of zinc dust and glacial acetic acid. The yield is about 40 %. Less readily proceeds the reduction with sodium amalgam in a weak acid solution. The acetate resulting from the zinc dust reduction is a liquid with a fruit-like odour, which boils at 98° to 102° (4 mm. pressure). From the acetic ester, the alcohol was obtained by saponification with alcoholic potash. It is a somewhat thick oil, rather difficult to volatilise with water vapours, of a pleasant, peculiar odour, and shows the following physical constants:

boiling point 89° to 91° (4 mm. pressure), 236° to 238° (755 mm. pressure); $d_{15°}$ 0,9419; $\alpha_D \pm 0°$; $n_{D 22°}$ 1,48652.

With phenyl isocyanate it forms a phenyl urethane of the melting point 100° to 101°, which crystallises from dilute alcohol in leaflets. The analysis of this derivative shows that the alcohol has the formula $C_{10}H_{18}O$:

0,1613 gm. of the substance: 0,4403 gm. CO_2, 0,1226 gm. H_2O.

	Found:	Calculated for $C_{17}H_{23}NO_2$:
C	74,46 %	74,70 %
H	8,44 %	8,44 %

This body is very stable towards strong formic acid at high temperatures.

The aldehyde $C_{10}H_{16}O$ discussed above, when left standing exposed to the air, readily oxidises into a leafy-crystalline acid $C_{10}H_{16}O_2$. This conversion proceeds more rapidly on treatment with moist silver oxide. The acid crystallising from petroleum ether or dilute alcohol melts at 106° to 107°, boils at 133° to 135° (4 mm. pressure), and is identic with the acid which had already previously[1]) been isolated by us from the saponification liquors of the oil.

Analysis: 0,1795 gm. of the subst.: 0,4706 gm. CO_2, 0,1544 gm. H_2O.

	Found:	Calculated for $C_{10}H_{16}O_2$:
C	71,50 %	71,43 %
H	9,56 %	9,52 %

---

[1]) Report April **1904**, 69.

Silver-determination: 0,2253 gm. of the substance: 0,0885 gm. Ag.

Found: Calculated for $C_{10}H_{15}AgO_2$:

Ag 39,29 % 39,27 %

The acid is capable of absorbing two atoms bromine, from which the conclusion may be drawn that the acid molecule, and consequently also the molecule of the aldehyde, shows a double-linking. According to this, and in view of the high specific gravity, the assumption is justified that the aldehyde is cyclic.

The brominating was carried out in chloroform solution, with ice-cooling. The reaction product crystallised from petroleum ether in leaflets with a silky lustre, of the melting point 116° to 117°. A bromine-determination proved that as a matter of fact 2 atoms bromine have been attached.

0,1830 gm. of the substance: 0,2104 gm. AgBr.

Found: Calculated for $C_{10}H_{16}Br_2O_2$:

Br 48,91 % 48,78 %

Oxidation of the aldehyde with 3 per cent. permanganate solution yielded as oxidation product a yellow syrup interspersed with minute crystals. This product has not yet been examined.

We are still occupied with the further examination of the alcohol $C_{10}H_{16}O$ and of the portions of the oil which have not yet been dealt with.

Gouft Oil. P. Jeancard and C. Satie[1]) obtained from two plants indigenous to Algeria, two new essential oils which they designated Essence de Gouft and Essence de Scheih. All parts of the plant were used up in the distillation.

The scheih oil had a brown-red colour and an odour like absinthe; the bright-yellow gouft oil reminded in the odour of turpentine and mastic.

The constants of gouft oil were as follows:

$d_{9,5°}$ 0,8720; a_D — 15° 20'; acid number 1,12; ester number 14; ester number after acetylation 42. The boiling point lay about 170°; the portions boiling below 170° appeared to contain pinene; melting point of the nitrosochloride 103°. From the higher-boiling fractions a few drops of an alcohol with a geraniol-like odour were isolated with phthalic acid anhydride.

Scheih oil had the following constants:

$d_{9,5°}$ 0,9540; acid number 8,4; ester number 66,5; ester number after acetylation 129,5. This oil contained obout 15 % phenols, among

---

[1]) Bull. Soc. chim. III. **31** (1904), 478.

which the principal constituent was found as dimethyl ether of pyrogallol. The oil freed from phenols distilled between 175° and 200°, and appeared to contain thujone and thujol.

Guaiacwood Oil. After a prolonged scarcity, a large consignment of wood was at last received in February, and this enabled us to execute all orders in hand. During the period of scarcity it has become evident what an important part this oil plays in the perfumery-trade.

Hop Oil. According to the reports received up to the present, the total result of the hop-harvest is on the average favourable. The copious rainfall in the beginning of September has worked wonders, especially in Bavaria, and even there a higher yield is now forthcoming than was originally expected. Increases as compared with last year are reported from Bohemia, Belgium, and America, whilst in England about the same crop has been picked.

The prices are lower, in agreement with the aforesaid conditions, but our distillation only commences within a few days, and to-day we are not yet able to quote a firm price for pure oil distilled from best non-sulphured Bavarian hops.

Jasmine "Schimmel & Co." The use of our original product is constantly growing. The preference shown for it is not surprising, in view of the many inferior competing products which are found on the market under the same name.

We have already mentioned previously that our product also contains a proportion of natural oil from the blossoms, which, jointly with the synthetic constituents, brings about the excellent effect of our oil. It is natural that such a perfect product possesses a real value, and cannot be sold at absurdly cut prices. For the rest, every consumer or seller himself can make the product cheaper by adding benzyl acetate.

In recent extraction-experiments with fresh jasmine-blossoms A. Hesse[1]) has obtained yields of oil which were twice as large as formerly. Accordingly, the enfleurage method would not give nine times, but only four to five times as much oil as the extraction process. At the same time, it must be assumed that a formation of essential oil from some not yet isolated substances takes place during the enfleurage period. Possibly the increased formation of odorous substances may be attributed in this as in other cases to fermentative disintegration of glucosides. In the extract obtained with petroleum

[1]) Berl. Berichte **37** (1904), 1457.

ether from fresh blossoms, no methyl ester of anthranilic acid could be detected, but this body was found in the essential oil obtained from this extract by steam distillation (0,4 %). From this it may be assumed that the compound yielding anthranilic acid ester is readily split up with water vapour. As the blossoms extracted to exhaustion with petroleum ether still yield comparatively large quantities of anthranilic acid ester when submitted to steam distillation, it follows that only a small portion of the compound is capable of extraction with petroleum ether. The substance forming indol is apparently not decomposed by steam distillation, for the distillation of fresh flowers again yielded oil free from indol.

Oil of Juniper berries. The harvest of berries both in Italy and in Hungary has given abundant results, and these two countries keenly compete with each other in the prices. In consequence of the dry weather, the berries have ripened well, and yield an excellent product.

Two oils of juniper berries originating from Russia, which have recently been examined in our laboratory, strangely showed a dextro-rotation of $+ 7°$ to $+ 8°$, whilst otherwise lævorotation is just a characteristic of oil of juniper berries. As the other properties corresponded with those of the ordinary distillate ($d_{15°}$ 0,8698; ester number 15,08; soluble in 5 and more vol. 90 per cent. alcohol), and as the oils also possessed a pleasant, powerful aroma, the deviation in the optical behaviour appears to be solely due to the origin of the oils. Further observations must show whether this assumption is correct, or whether perhaps some other vegetable material serves for the production of this oil.

Oil of Laurel leaves. In our last Report[1]) in discussing Molle's thesis, we have already referred to the work published by H. Thoms and B. Molle[2]) on the composition of the essential oil of laurel leaves, and for this reason we need now only refer to that Report.

Lavender Oil. For this eminently important article, the abnormal drought during the summer-months has also had disastrous results: on this point all reports from the leading departments agree, so that it will not be necessary to make special mention of the results in the individual districts. One of our principal purveyors estimates the harvest at one-fourth of a normal one, and states that many portable stills could not be worked for want of cooling-water. There is moreover lack of labour for picking the half-dried up plants.

---

[1]) Report April **1904**, 54.
[2]) Arch. der Pharm. **242** (1904), 161.

The quantity offered at the first market at Sault was correspondingly unimportant. Oils containing 33 to 35 % ester were bought by firms in the South of France at 18 francs. At the markets subsequently held at Digne and Apt, oils with 40 % ester-content fetched 22 francs. The inclination to buy was exceptionally strong, — a fact which unfortunately justifies the view that no stocks of any importance are held by firms in the South of France. A definite review of the whole situation is not yet possible, but so much is certain, that the quantity produced falls far short of that of last year, and that for the next twelve months very high prices will have to be reckoned with.

Lavender oil is one of the specialities of our firm. We have arrangements for the most advantageous purchases, and we are kept informed in the best possible manner of everything that passes at the principal centres.

Under the critical conditions described above, the adulteration of lavender oil plays an important part, and there is every indication that it will be carried on this year on an extensive scale. We learn from a private source that at the places of production spike oil comes chiefly under consideration as the adulterant, and further turpentine oil and methylated spirits. Spike oil is either added to pure lavender oil, or the adulteration already takes place in the still by submitting a mixture of spike- and lavender-blossoms to distillation. The latter occurs chiefly in the lower regions, where spike and lavender grow side by side. In order to lessen the camphor-odour due to the spike oil, the oils are still washed with water. But the more skilled adulterators proceed in a decidedly more cunning manner. They use for their "experiments" Spanish oil of sage, which can be bought in France at a low price; to this they give the requisite solubility by adding spike oil, and the necessary aroma by a few drops amyl acetate.

According to the information from our authority, almost every lavender oil which contains less than 33 % ester is adulterated. There are of course exceptions, but they are few; for example, the oils from the left bank of the Var are greatly valued by many purchasers, although they contain only 28 to 30 % ester. These oils are characterised by a very low specific gravity and a fairly high lævorotation, whilst their solubility is normal.

Our own experience absolutely agrees with the above information relating to the adulteration of lavender oil; we also have frequently had an opportunity of observing that spike oil plays an important part in this matter. Two lavender oils recently examined by us are also interesting for this reason, that to all appearances petroleum, or a petroleum-fraction, had also been used as an adulterant. The two

4*

oils which clearly originated from the same source, were entirely identic: $d_{15°}$ 0,8707; a_D — 3° 22'; ester-content 21,26 %; insoluble in 10 vol. 70 per cent. alcohol.

Unfortunately the quantity at our disposal was too small for an exact examination, and the nature of the adulterant could therefore not be determined with certainty. But if our above-mentioned assumption should be confirmed, one more adulterant would be added to the number which comes under consideration for lavender oil.

Lemongrass Oil. We have on several occassions reported on the experiments which have been made in the Botanical Garden of Victoria (Cameroons) with the cultivation of andropogon grasses for the purpose of oil manufacture[1]). Whilst the oil distilled from these grasses was considered by Strunk, the Director of the Botanical Garden, to be citronella oil, Mannich found, from a sample sent to the Pharmaceutical Institute of Berlin, that the oil is identic with lemongrass oil. A sample which we recently received from Dr. Strunk gave a result which corresponded with that obtained by Mannich. The oil had a yellow colour and a powerful, pleasant aroma: $d_{15°}$ 0,8929; a_D — 0° 8'; aldehyde-content (citral) 84 %; the solubility is imperfect, inasmuch as the oil does not at all form a clear solution with 70 per cent. alcohol, and the solutions in 80 and 90 per cent. and even absolute alcohol, which are at first clear, also become cloudy with more or less strong opalescence, when more alcohol is added. The oil resembles on this point West Indian lemongrass oil, and is as little as the latter able to replace the East Indian oil, owing to its imperfect solubility. But, on account of its high aldehyde-content, it may possibly be found suitable as crude material for the production of citral.

On a previous occasion[2]) we have already pointed out that the andropogon grasses, where their origin is mentioned, are frequently mistaken one for the other. This inconvenience is all the more felt, as the uncertainty of the botanical nomenclature also exists in scientific works.

It is as yet an open question whether — as is frequently believed — the climate and character of the soil exert such an influence on the plants, that one and the same species under changed conditions yields oils of which the composition differs considerably; but there is a certain amount of probability in favour of this belief, and it would also partly explain the above-mentioned inconvenience. We have repeatedly had before us authentic oils originating from the West Indies, which according to the information received by us had been derived from *Andro-*

[1]) Report April **1903**, 23; October **1903**, 46.
[2]) Report April **1903**, 23.

pogon Schoenanthus, and which therefore (in agreement with the East Indian distillates) should have been palmarosa oils, but which had a much greater resemblance to lemongrass oil. Similar observations have also been made in the laboratory of Roure-Bertrand Fils at Grasse[1]), and the view here represented, that the same plant yields in one case an oil rich in geraniol, and under other conditions produces an oil which contains in a preponderating degree the corresponding aldehyde citral, has a great deal of probability in its favour.

Ch. J. Sawer now has set himself the task of clearing up the existing contradictions as much as possible by a thorough botanical study of the andropogon grasses, with simultaneous reference to the literature on the subject, — an enterprise worthy of acknowledgment, which has unfortunately been suddenly interrupted by Sawer's death in August last.

It is seen from Sawer's work[2]) that the solution of the question under discussion is attended with great difficulties, as the andropogon grasses are plants which show fairly large variations, which renders a differentiation between the individual species much more difficult. This applies especially to *Andropogon Schoenanthus* L. To this should be added that many sub-species, varieties, and also forms of transition are in existence. In consequence of this, confusion of the individual species may easily occur, but on the other hand, the oils yielded by the different varieties also show differences in their properties and composition, which explains why it is not always possible to draw a definite conclusion from the oil as to its mother-plant. Moreover, climate and nature of the soil, and (especially in the case of citronella oils), the time at which the grass is cut, are said to have an important influence on the oil.

Sawer next describes the individual kinds of grass, and discusses citronella grass and lemongrass. This part amounts in the main to a classification of the botanical literature dealing with these grasses, and for this reason we refrain from entering into it in detail. For greater clearness a number of characteristic illustrations are added. But Sawer's work shows at any rate that among the botanists who have made a special study of the andropogon grasses, differences of opinion on the nomenclature of the individual mother-plants also exist. Moreover, several grasses have not yet been determined at all with certainty from a botanical point of view, as for example Indian citronella grass, West Indian lemongrass, etc.

It may be interesting to mention here that in Java lemongrass oil is called *sireh*, but the same name is given, there to the Javanese

---

[1]) Report of Roure-Bertrand Fils, Grasse, April **1904**, 40.

[2]) Citronella and Lemongrass. Chemist and Druggist **65** (1904), 179. Comp. also the publication by J. C. Willis relating to this subject, ib. 478.

herb *Tetranthera citrata* which has a very similar odour. In Mexico lemongrass is called *Te limón*, as we have already mentioned in a previous Report[1]. According to Sawer a beverage prepared from lemongrass is called there by the same name.

Limette Oil. To the kindness of Sir Daniel Morris, Imperial Commissioner of Agriculture in Barbadoes (W. I.) we are indebted for two limette oils[2]) originating from Dominica, viz., one obtained by expression (*hand-pressed lime oil*), and a distilled oil (*distilled lime oil*). We give below the constants of the two oils, which, in view of the comparative scarcity of authentic samples, are of particular interest.

1. *Hand-pressed lime oil:* $d_{15°}$ 0,9008; $\alpha_D + 36° 17'$; α_D of the first 10 % of the distillate $+ 39° 30'$; acid number 6,05; ester number 29,55; residue on evaporation 17,8 %; soluble in 4 and more vol. 90 per cent. alcohol with slight cloudiness in consequence of separation of paraffin. The dilute alcoholic solution shows a faint blue fluorescence, which renders the presence of anthranilic ester probable. The oil has a golden yellow colour and possesses a pleasant refreshing odour which greatly resembles that of lemon oil.

2. *Distilled lime oil.* The bright-yellow oil, which possesses a disagreeable odour like turpentine or pine tar oil, had the following constants: $d_{15°}$ 0,8656; $\alpha_D + 46° 36'$; α_D of the first 10 % of the distillate $+ 53° 8'$; acid number 1,8; ester number 4,05; residue on evaporation 3,16 %; soluble with slight cloudiness in 4,5 and more vol. 90 per cent. alcohol.

H. E. Burgess and Th. H. Page[3]) report on the composition of distilled limette oil. For the examination they used a concentrated oil, which by distillation at reduced pressure was divided into two principal fractions, of which the former, passing over, at 17 mm., between 100° and 105°, consisted chiefly of l-terpineol of the melting point 35°. The liquid portions were removed from it by freezing out, and it was then purified by recrystallisation from alcohol: $d_{15°}$ 0,941; $\alpha_D - 20°$; $n_{D 20°}$ 1,4829; boiling point 214° (762 mm.). The compound was further identified by the production of various derivatives.

As the fraction containing the terpineol had the characteristic odour of distilled limette oil, but the isolated terpineol of the melting point 35° is almost odourless, Burgess and Page believe that the actual odour-

---

[1]) Report October **1903**, 46.
[2]) With regard to limette oil, comp. also Gildemeister and Hoffmann, "The Volatile Oils", p. 477.
[3]) Journ. chem. Soc. **85** (1904), 414. Unfortunately the authors do not state the origin of the oil, a matter which is rather important in view of the difference between the West Indian and Italian oils. We surmise that it was a West Indian oil.

bearer must be found in an isomeric liquid terpineol of a somewhat lower boiling point, but they were unable to produce any proof in support of this.

The other fraction boiling from $130°$ to $140°$ (17 mm.) consisted almost entirely of hydrocarbons. The authors succeeded in isolating from it a sesquiterpene $C_{15}H_{24}$ belonging to the class of light sesquiterpenes, to which they have given the name "limene". It represents a colourless oil, with a faint but peculiar odour, which readily resinifies, and boils at .9 mm. at $131°$, whilst at ordinary pressure (756 mm.) it passes over between $262°$ and $263°$ with slight decomposition. The other properties of limene are: $d_{15°}$ 0,873; $a^D \pm 0°$; $n_{D\,19,5°}$ 1,4910; molecular refraction 68,2, calculated for $C_{15}H_{24}$ (with three double-linkings) 67,76; with this agrees the fact that the compound absorbs 6 atoms bromine.

Of the derivatives, only the trihydrochloride $C_{15}H_{24}$. 3 HCl could be produced, with the help of which the hydrocarbon had also been isolated. It is characterised by a strong capacity of crystallising, and forms colourless crystals of the melting point $79°$ to $80°$, which readily dissolve in ether, acetic ether, and acetone, but are less soluble in alcohol, glacial acetic acid, and chloroform.

Limene was also found in limette oil obtained by expression, and further in lemon oil; other oils will still be examined for it.

We would point out that we also isolated some time ago from terpeneless lemon oil a light sesquiterpene[1]) which agrees well with the one described above, except in the rotation (a_D — $42°$). The difference in the optical behaviour may possibly be explained by the fact that Burgess and Page accomplished the production in the pure state by means of the hydrochloride.

Linaloe Oil. The value of this article during the last six months did not undergo any pronounced fluctuations. It is said that improved installations for the distillation of linaloe oil have recently been erected in Mexican territory by a certain party, which on the spot where the wood is felled works up exclusively the core of centurial linaloe-trees, stated to yield an oil of special quality. From the same quarter it is somewhat rashly reported that the distillation will in a few year's time come to a standstill owing to lack of wood. Such statements do not tend to promote sympathetic feeling towards this undertaking, for as far as we know, the linaloe-tree grows in different Mexican provinces. The oil distilled in French Guyana from so-called *Bois de rose femelle* is now exceedingly well introduced in commerce; a large trade is, however, prevented by the high prices.

As a distinctive sign of the purity of linaloe oil, lævorotation is taken among others; this should usually not be less than $5°$. Partial

---

[1]) Report April **1903**, 37.

removal of the linalool, or addition of waste products which originate from the manufacture of linalool from linaloe oil, exert a rather large influence on the rotation of the oil, which not infrequently changes into dextrorotation, whilst the other properties are but slightly altered. But the saponification number which in normal oil lies between 1 and 2 5, generally increases then up to 30—45, and thus also becomes a betrayer of the adulteration. In such case fractional distillation *in vacuo* will readily show that the oil is poor in linalool, and therefore adulterated.

On the other hand, dextrorotation can also be often observed in linaloe oil, if the individual fractions collected during the distillation are not mixed equally. Fractional distillation will here also give the required information, at any rate l-linalool can always be isolated from such an oil.

For this reason the observation made in our laboratory is all the more interesting, that dextrorotatory linaloe oil of otherwise good quality also exists.

The oil examined had the following constants: d_{15° 0,8793; $a_D + 7^\circ 31'$, acid number 1,02; ester number 3,88; soluble in 1,7 and more vol. 70 per cent. alcohol.

This oil differs from the above-described abnormal oil by its low ester number, and the further examination showed that the oil was quite normal. On fractional distillation a fraction boiling between 94° and 98° (13 mm.) was obtained, which represented about 65 % of the oil, and which was identified on further examination as d-linalool: d_{15° 0,8701; $a_D + 11^\circ 15'$; n_{D20° 1,46209. The phenyl urethane[1]. produced from it melted at 65°; oxidation with chromic acid mixture yielded citral.

Unfortunately we have been unable to ascertain anything more definite about the origin of the oil, and we must therefore content ourselves with mentioning the observation made by us.

Matico Oil. Some time ago Fromm and van Emster[2]) reported on the composition of matico oil. The authors isolated from the heavy portions of an oil distilled by us, as principal constituent, an unsaturated phenol ether "matico ether", to which they ascribed the formula of a methylbutenyl-dimethoxy-methylenedioxybenzene. On oxidation matico ether yielded an aldehyde and also an acid; with regard to the relative situation of the substitutes in the benzene-nucleus nothing was stated. Thoms[3]) took up the examination of

---

[1]) The linalylphenyl urethane, in 10 per cent. alcoholic solution, showed a rotation of $+ 1^\circ 6'$. As a trial showed that the phenyl urethanes rotate in the same way as the corresponding linalools, this also proves that we had here to deal with d-linalool.

[2]) Berl. Berichte **35** (1902), 4347, — Report April **1903**, 51.

[3]) Arch. der Pharm. **242** (1904), 328.

the heavy matico oil and found in it, in addition to matico ether, a hydrocarbon congealing at — 18°, and a further phenol ether. The so-called matico ether, as Thoms was able to prove, is composed of two apiols, namely parsley apiol

$$CH_2:CH \cdot CH \underset{H \quad OCH_3}{\overset{OCH_3 \quad O \longrightarrow CH_2}{\langle \qquad \rangle}} O$$

in a small proportion, and dill apiol

$$CH_2:CH \cdot CH \underset{OCH_3 \quad OCH_3}{\overset{H \quad O \longrightarrow CH_2}{\langle \qquad \rangle}} O$$

as principal constituent. In agreement with the above, two isomeric apiolic acids of the melting point 175° and 151° respectively occurred on oxidation. The identification of parsley apiol was perfected by the reduction of the isomerised phenol ether with sodium and alcohol, methylation of the phenol thereby obtained, and nitration of the ether thus formed; the nitro-product was found to be identic with the previously obtained 1-propyl-2, 3, 5-trimethoxy-4-nitrobenzene. Fromm confirms the results of this examination in a postscript.

Mignonette (Reseda) Geraniol. The mignonette-plants which owing to an indescribable drought have only flowered most poorly, have given such a small yield that the production of mignonette geraniol has been very limited. It was unfortunately necessary to reduce the allotted quantities very much. However annoying this is, it should be borne in mind that one is absolutely powerless against such abnormal influences of the weather.

Oil of Monarda citriodora. An examination of the essential oil of *Monarda citriodora*, made by J. W. Brandel[1]), has shown the following results: the dried herb contained 1% of a reddish oil of the specific gravity 0,9437 (20°), 65% of this oil consisting of phenols, of which carvacrol was identified by the benzoyl compound of the nitroso derivative (melting point 110°), and hydrothymoquinone by its melting point 140°. Citral was found in the oil in a quantity of 1,2%, determined by Sadtler's method. In a fraction of the oil freed from phenol, boiling at 170° to 175°, the author believes to have detected cymene.

Oil of Monarda didyma. In the Swiss pharmaceutical collection of popular medicines a vegetable drug is found according to

---

[1]) Pharm. Review **22** (1904), 153.

J. G. Gerock[1]), which under the name of "Goldmelisse" is largely used in the district about Berne. Under this name is known *Monarda didyma* L. which is also indigenous to Southern Canada and the mountain-ranges of the State of Georgia. The plant has a powerful aromatic odour (which, however, is not at all like that of balm), and it owes its stimulating and diuretic action undoubtedly to the essential oil contained in the leaves and blossoms. In America the plant is also known under the popular name of beebalm or horsemint, and is used there for the same purposes as in Switzerland as Oswego tea or Pennsylvania tea. The composition of the essential oil *Monarda didyma* is still unknown. J. W. Brandel[2]) states that the plant yields $0,03\,^0/_0$ essential oil, and, contrary to earlier statements, does not contain either thymol or carvacrol in demonstrable quantities.

Oil of Monarda fistulosa. Experiments made by F. Rabak[3]) have proved that the thymoquinone[4]) occurring in the essential oil of *Monarda fistulosa* in addition to hydrothymoquinone, must be attributed to the presence of an oxidising ferment. Rabak isolated it from the plant by bruising the fresh leaves in a mortar and stirring them with water into a pulpy mass. From the expressed filtered liquid the ferment separated out when alcohol was added. This precipitate apparently only reacts with hydrogen peroxide, whereas with tincture of guaiac its aqueous solution already produces a deep-blue colour. The author was able to demonstrate by special experiments that the oxydase reacts on the hydrothymoquinone first of all with separation of dark crystals which are gradually converted into the yellow aggregates of thymoquinone. This process may probably be explained in this manner, that the particles of thymoquinone first formed produce with as yet unchanged hydrothymoquinone, thymoquinhydrone, which then by the continued action of the oxydase yields thymoquinone. In an exactly analogous manner the ferment converts hydroquinone into quinhydrone; but a further oxidation into quinone does not seem to take place. Nor can an oxidising action on carvacrol or thymol be observed, nor on cymene which stands in close relation to these two phenols. The author, in consideration of the fact that in a mint an oxidising ferment was found, also allowed the Monarda oxydase to act on menthol, but here also he could only observe a negative result.

D. B. Swingle[5]) has subsequently made further experiments with this ferment, and on the strength of the results obtained by him has

---

[1]) Journ. der Pharm. von Elsaß-Lothringen **31** (1904), 78.
[2]) Pharmaceut. Review **21** (1903), 109. Comp. Report Oct. **1903**, 51.
[3]) Pharmaceut. Review **22** (1904), 190.
[4]) Pharmaceut. Review **19** (1901), 200, 244, Report Oct. **1901**, 73.
[5]) Pharmaceut. Review **22** (1904), 193.

expressed the supposition that the soluble Monarda ferment may be identic with β-*katalase* of Löw[1]). The temperature at which the ferment is destroyed lies about 74° to 78°.

Mustard Oil, genuine. The material was supplied in the first place by India, next by Russia, which latter country also placed large parcels of mustard-seed cakes freed from fatty oil on the market. It is well known that the fatty oil is used in Russia for purposes of nutrition. In the Italian province of Puglia, according to a report from the German Consul at Bari, the mustard production diminishes from year to year owing to the competition from India. Moreover, the harvest was injured by heavy rainfall. Dutch mustard, which for many decades has served chiefly as material for mustard oil, is now out of the question for this purpose as the price is too high.

Artificial mustard oil has completely corrupted the trade in genuine oil, and offers are now met with, which almost approach the prices of artificial oil. Whoever buys at such prices in the belief of purchasing genuine oil, commits a first class act of self-deception.

To the proposals for modifying the directions for the determination of mustard oil mentioned in our last Report[2]), is now added one by Vuillemin[3]) who occupied himself specially with K. Dieterich's estimation-method. The difference between the working methods lies chiefly only in the temperature of the water employed, and in the duration of the action; the author also recommends the addition of a small quantity of alcohol to the ammonia present in the receivers, in order to prevent the passing over of the mustard oil into the second receiver. Vuillemin prefers the factor 0,4301 instead of 0,4311, in view of the fluctuating content of carbon disulphide, cyanallyl, and sosulphociyanallyl in the oil. The determination of mustard oil in mustard seed is accordingly carried out in the following manner: —

5,0 gm. mustard seed are triturated as finely as possible, placed in a round flask of 200 cc. capacity, mixed with 100 cc. tepid water (25° to 30°), and left standing well closed for an hour, with frequent agitation. 20 cc. alcohol are then added, the flask connected with a Liebig's condenser, an Erlenmeyer flask of 200 cc. capacity with 30 cc. ammonia liquor and 10 cc. alcohol, and about half the contents distilled over whilst the condensing tube is immersed in the liquid. The Erlenmeyer flask used as condenser is connected with a second flask containing ammonia liquor and alcohol, in order to prevent any

---

[1]) U. S. Dept. of Agr. Report No. 68, 7.
[2]) Report April **1904**, 63.
[3]) Pharm. Centralh. **45** (1904), 384.

loss whatever. The condenser is rinsed with a little water, the distillate mixed with 3 to 4 cc. solution of silver nitrate (1 : 10) and heated on a water-bath until the conglomeration of silver sulphide is well deposited, and the liquid is absolutely water-white. The precipitate is collected by filtering the hot liquid on a chemically pure filter of 5 to 8 cm. diameter, the flask and precipitate washed one after the other with a small quantity of hot water, alcohol and ether, and dried at 80° C. until its weight remains constant. The silver sulphide thus obtained, when multiplied with 8,602, indicates the percentage of mustard oil in the seed examined.

Mustard oil, artificial. Our plant for the manufacture of this oil has been fully employed. The consumption was normal; the prices are unchanged.

Neroli oil, French. The blossom-harvest in the department Alpes-Maritimes has been fairly normal this year; the yield is estimated at about 1 800 000 kilos. The oil-content of the blossoms was also favourable. For this reason a movement to increase the prices of neroli oil, which was set on foot from various quarters, was all the more surprising.

This increase is a consequence of the acute differences which have broken out between the manufacturers and the growers, leading to the formation of a syndicate of the latter which controls about two-thirds of the entire quantity of blossoms. The object of this syndicate is to raise the price of the blossoms, which has been continually depressed by the manufacturers and has become ruinous for the growers, up to 65 francs per 100 kilos for the whole season, and if this price cannot be obtained, the syndicate itself will distil the blossoms. It is even said that the syndicate has destroyed about 40 000 kilos blossoms which could not be distilled this year owing to lack of apparatus. For the next season, arrangements would be made for working up the whole quantity of blossoms of the syndicate. It is further stated that the syndicate has distilled this year about 300 kilos neroli oil, but it may possibly not be such a simple matter to dispose of this oil at the comparatively high price asked, the less so, as firms outside the syndicate are offering at lower prices. The question is now: Will the syndicate succeed in bringing the majority of the blossom-producers under one flag, and keeping them together? Experience teaches that as a rule such trusts come to grief owing to internal quarrels.

It must be admitted that the prices which the manufacturers paid for the blossoms frequently barely covered the cost of picking, and from this point of view one might perhaps wish success to the "Société coopérative des Propriétaires de Fleurs d'oranger".

The movement is also supported by the press; "Le Petit Marseillais"
publishes the following article in its issue of June 7th: —

La question de la fleur d'oranger.

On nous écrit d'Antibes, le 6 juin:

La cueillette de la fleur d'oranger est terminée et celle de la montagne
le sera sous peu. Il a fallu dix-huit jours pour la clôturer, alors que les
années précédentes elle durait de vingt-cinq à trente jours; rien ne prouve
mieux l'état d'une faible récolte. Malgré cette pénurie, les transactions ont été
laborieuses à cause des hostilités préconçues de MM. les parfumeurs.

Heureusement les préposés à la vente de la fleur étaient des hommes
choisis, énergiques et expérimentés, qui ont pu s'acquitter de leur mission à
l'avantage de la Société.

MM. les fondateurs peuvent être fiers de leur œuvre, mais c'est surtout
à M. Gagnaire, notre estimé président, à qui revient la plus grande somme
d'honneur. Il vient de prouver qu'il n'y a d'obstacle qui ne puisse être sur-
monté avec de la persévérance et de la suite dans une idée. Nous l'en félicitons.

Nous pouvons assurer dès à présent, sans crainte d'exagération, qu'à la
récolte prochaine les résultats seront encore meilleurs, attendu que nous serons
alors en mesure de mieux nous défendre au cas où MM. les parfumeurs per-
sisteraient à ne pas vouloir entrer en relation d'affaires avec nous.

Nous ne voudrions pas terminer cette chronique sans adresser quelques
mots aux propriétaires de fleurs libres et aux abonnés; aux premiers nous
dirons: vous avez été pusillanimes et incrédules, car vous avez douté de la
solidarité, qui crée la fraternité; vous n'avez pas eu foi en l'union qui fait la
force, mais vous n'avez pas cru à la toute-puissance des parfumeurs et vous
avez eu peur, alors vous vous êtes tenus prudemment à l'écart en spectateurs
pour juger du coup et voir qui l'emporterait. Aujourd'hui l'expérience est faite,
elle est concluante et malgré les douze plus puissants parfumeurs qui ont persisté
à ne pas vouloir acheter un kilo de fleurs à la Société.

Les 1 200 000 kilos dont elle disposait ont été écoulés à des prix rému-
nérateurs.

D'ailleurs à la récolte prochaine des usines créées par la Société seront
prêtes à fonctionner, dans le cas probable où MM. les parfumeurs renouvelleraient
leurs promesses de cette année. Dans ces conditions vous ne devez plus avoir
à craindre d'incertitude et ne plus hésiter à venir à nous, aux abonnés anciens
et nouveaux nous dirons à l'expiration de ce mandat: Ne le renouvelez plus,
ne donnez plus au parfumeur l'arme pour nous battre, que l'égoïsme n'étouffe
pas en vous le sentiment de la solidarité. Venez à nous, venez grossir nos
rangs et dites-vous bien que le jour où tous les propriétaires d'orangers seront
syndiqués, l'âge d'or reviendra pour eux; car alors la fleur ne sera plus payée
au-dessous de 60 centimes.

In the above-described question at issue between the parties, the
presence of artificial neroli oil might prove useful to the manufacturers, for it is now recognised on all hands that a product like
ours is at least equal to the natural oil, although it stands to reason
that in the South of France appreciation is withheld as a matter of
principle. It is to be hoped that the increased prices of natural oil
will act as a fresh stimulant to thorough comparative trials.

P. Freundler[1]) has discovered a suitable reaction for the detection, and possibly also for the quantitative estimation, of methyl ester of anthranilic acid. In a manner similar to that in which H. N. Mac Coy[2]) obtained thiophenyl ketotetrahydroquinazoline from anthranilic acid by heating it with phenyl mustard oil and alcoholic soda liquor, Freundler obtained this compound in a quantitative yield, simply by heating methyl ester of anthranilic acid with phenyl mustard oil to 100° to 120°. The conversion takes place according to the following equation:

$$C_8H_4 \underset{NH_2}{\overset{COOCH_3}{<}} + S:C:N \cdot C_6H_5 = C_6H_4 \underset{NH}{\overset{CO-N \cdot C_6H_5}{<}} CS + CH_3OH.$$

The compound melts above 300°, is readily soluble in soda liquor and very difficultly in alcohol. From 5,1 gm. pure methyl ester of anthranilic acid, 98% of the calculated quantity of quinazoline was obtained. With regard to the application of the method for the determination of methyl ester of anthranilic acid in essential oils, for which the authors propose it, no experiments are as yet mentioned. But Freundler used this derivative to confirm the statements made by E. von Meyer and M. Schmidt[3]) that on heating isatoic acid with methyl alcohol to 130°, methyl ester of anthranilic acid is formed. E. and H. Erdmann[4]), on applying for a patent for the manufacture of this compound, had contested the correctness of the result obtained by E. von Meyer and M. Schmidt, and consequently the latter's priority for the synthesis of this ester.

When repeating the experiment under the conditions mentioned by E. von Meyer and M. Schmidt, E. Erdmann[5]) was, however, compelled to admit that as a matter of fact methyl ester of anthranilic acid is formed. Whilst Freundler employed phenyl mustard oil for the separation of methyl ester of anthranilic acid from the other products formed by the action of isatoic acid on methyl alcohol, he established that at least 28% of the isatoic acid employed had been converted into methyl ester of anthranilic acid. We, who by the detection of methyl ester of anthranilic acid in neroli oil and other blossom oils, and also by the technical application of the synthetic ester were the first to demonstrate the importance of the compound as an odorous substance, have from the beginning admitted the identity of that body

[1]) Bull. Soc. Chim. III. 31 (1904), 882.
[2]) Berl. Berichte 30 (1897), 1688.
[3]) Journ. f. prakt. Chem. II. 36 (1887), 374.
[4]) German Patent No. 110386, cl. 12; March 16, 1900.
[5]) Berl. Berichte 32 (1899), 2168.

with the one obtained by E. von Meyer and M. Schmidt[1]). The view then expressed by us finds a welcome confirmation in the results of Freundler's examination.

In addition to thiophenyl ketotetrahydroquinazoline, Freundler has also produced the picrate of anthranilic acid ester not yet described. It is formed when alcoholic solution of picric acid is mixed with methyl ester of anthranilic acid, and it crystallises in yellow needles which dissolve fairly readily, and have the melting point 103° to 104°[2]).

In continuation of their work[3]) on the formation and distribution of organic substances in the plant, E. Charabot and G. Laloue[4]) have now studied the production of essential oil in the orange tree during its period of vegetation (as previously with the geranium plant and the mandarin tree). They observed that the formation of the oil proceeds most briskly at the beginning of the vegetation-period. At this stage the oil in the young leaves is not so rich in esters and total alcohols as that of the young twigs and stalks. Towards the end of the development of the plant this difference in the ester-content becomes even more pronounced. The oil of the old leaves is then however much richer in total alcohols than that of the stalks, — especially in linalool, less so in geraniol. Between these two stages of development the leaf-oil experiences a very pronounced decrease in its content of linalool, and against this a slight enrichment in esters. The oil in the stalks on the other hand becomes in a marked degree richer in esters, but strikingly poorer in total alcohols. It follows from the foregoing that the oil in the stalks is less soluble than that in the leaves. In consequence of this, the osmotic pressure in the stalks, according to the law of diffusion, will constantly decrease, and there-fore a transition of a portion of the oil will take place from the leaf towards the stalk, with displacement of a portion of the less soluble oil from the saturated solution in the stalk. A second work by the above-named authors[5]) deals with the distribution of the essential oil in the orange blossom. According to this, of all blossom organs the leaves of the blossom contain the largest quantity of oil. As a matter of fact, the oil-content of the blossom decreases considerably during the flowering-period. At this stage, the production of the odorous substances is much more brisk than at any earlier period. In the course of the development of the blossoms an enrichment of the oil

---

[1]) Journ. f. prakt. Chem. II. **59** (1899), 352.
[2]) According to our observations the melting point of this compound lies at 105° to 106°.
[3]) Report October **1903**, 40; Report April **1904**, 50.
[4]) Compt. rend. **138** (1904), 1228. Bull. Soc. Chim. III. **31** (1904), 884.
[5]) Compt. rend. **138** (1904), 1513. Bull. Soc. Chim. III. **31** (1904), 937.

in esters of terpene alcohols, methyl ester of anthranilic acid, and total alcohols takes place. The relation between the quantities of esterified alcohols and of total alcohols increases, from which it may be concluded that the ester-formation in the blossom continues, though only slowly. It is here, however, less complete than in the leaf and stalk. The geraniol-content increases, but that of linalool diminishes. The consequence of this is, that the mixture of alcohols becomes richer in geraniol. After the blossom is unfolded no pronounced differences between the essential oil of the leaves of the flower and that of the other blossom organs can be detected; the former only contains a somewhat larger quantity of methyl ester of anthranilic acid than the latter.

Nigella Oil. H. Pommerehne[1]) and O. Keller[2]) communicate some further results of their examination of damascenine. The former, on oxidising damascenine hydrochloride with barium permanganate, obtained in addition to oxalic acid, also volatile bases consisting of ammonia and methyl amine; he further found that when damascenine is heated with baryta water, it shows the same behaviour as when boiled with more strongly alkaline liquids. This base, namely, as we have already shown, is converted quantitatively during boiling with alcoholic potash into a beautifully, crystallising amido acid which, according to Pommerehne's examination, is isomeric with damascenine, and is also formed from the latter on treatment with sodium carbonate[3]).

O. Keller has occupied himself with a closer examination of this acid, to which provisionally the name damascenine-S has been given. Damascenine-S crystallises from water in rhombic prisms with three molecules water of crystallisation. The crystals which effloresce rather quickly when exposed to the air, melt at $78°$, and in alcoholic solution show a beautiful blue fluorescence. Dried at $90°$, the anhydrous acid melts at $143°$ to $144°$.

From the analysis the formula $C_9 H_{11} N O_3 + 3 H_2 O$ is calculated.

When the molecular weight was determined, the figures obtained corresponded to the values calculated for the above formula. When bromine acts on damascenine, the hydrobromate of dibromdamascenine is formed, which proves the presence of a double-linking in damascenine. This double-linking is also present in the conversion-product of the bases, damascenine-S, for the dibromide $C_9 H_{11} Br_2 N O_3$ was also obtained from the latter. By boiling with acetic acid anhydride Keller had produced from damascenine a monoacetyl product of

---

[1]) Arch. der Pharm. **242** (1904), 295.
[2]) Arch. der Pharm. **242** (1904), 299.
[3]) Report October **1899**, 40. Pommerehne, Arch. der Pharm. **238** (1900), 546.

the melting point 203° to 204°; a substance of the same composition and the same melting point was formed when damascenine-S was treated in the same manner. Consequently damascenine is also converted into the isomeric amido acid when boiled with acetic acid anhydride. No new compound was obtained when methyl iodide was allowed to act on the acetyl product. But Pommerehne had already found that the isomeric amido acid from damascenine reacts with methyl iodide.

Keller's examination showed that an addition-product

$$C_9H_{10}NO_3 \cdot CH_3 \cdot HI + H_2O$$

of the melting point 172° to 173°, is here formed, which is identic with the one obtained direct from damascenine. Damascenine, therefore, is also · converted by methyl ioide into damascenine-S. From the above compound the new base $C_9H_{10}O_3 \cdot N \cdot CH_8$ (melting point 118° to 119°) was isolated with sodium carbonate; this base again absorbs one molecule methyl iodide with formation of a compound

$$C_9H_{10}O_3 \cdot N(CH_3) \cdot CH_3I + H_2O,$$

which crystallises with one molecule water.

Both damascenine-S, and the isomeric amido acids, when treated with sodium nitrite in hydrochloric solution, yielded the same nitroso acid $C_9H_{10}O_3N \cdot NO$ of the melting point 151° to 152°.

When damascenine-S was heated for two hours with hydriodic acid of the boiling point 127° and phosphorus to 150°, an amido phenol was formed which melts at 170°, and which is therefore identic with o-amido-phenol. From these results the following formula may be deduced for damascenine-S: —

$$C_6H_3 \diagdown \begin{matrix} OCH_3 \\ NH \cdot CH_3 \\ COOH \end{matrix}$$

in which the methylated amido-group is in ortho-position towards the methoxyl-group. For this reason the isomeric amido acid formed from damascenine is nothing but an o-anisidine-carboxylic acid methylated in the amido-group.

Nutmeg Oil. In a nutmeg oil recently distilled by us, a higher specific gravity and lower rotatory power were determined than usually. Whereas the properties of nutmeg oil are generally as follows: $d_{15°}$ 0,870 to 0,920; $a_D + 11°$ to $+ 30°$; soluble in 1 to 3 vol. 90 per cent. alcohol, — the oil in question had a specific gravity of 0,9220 at 15°, and an optical rotation of $+ 7° 52'$; moreover it was already soluble in 0,5 vol. 90 per cent. alcohol. These differences must be attributed solely to the quality of the nutmeg worked up; in the present case the quality was particularly good, contrary to the usual distillation quality which consists chiefly of light worm-eaten nuts. The oils distilled from the last-named material are consequently richer in terpenes, whilst in the other oil more oxygenated constituents are present. This may sufficiently explain the difference obtained.

On this occasion we would mention that we sometimes receive complaints from the United Kingdom on account of too high specific gravities of our oils of nutmeg, which in consequence do not answer the requirements of the British Pharmacopœia. The B. P. gives as limits of the specific gravity 0,870 to 0,910 (15,5°), and thereby, — as on many other points relating to essential oils — takes no account of the actual facts; it really excludes from medicinal use the oils produced from the best nutmeg. A radical elimination of such contradictions which unfortunately do not stand alone, is greatly to be desired.

A long treatise by Gillavry[1]) deals with the cultivation of the nutmeg trees in Djati Roengge (Java). When the author in 1876 came to Djati Roengge, he only found a 45 year old plantation of 180 nutmeg trees, which to-day are still in existence and bear fruit. In the following year 5000 Banda nuts were planted, and the plants obtained from these seeds transplanted only in 1879, when they were about 18 inches high and possessed a sufficiently developed root-system. Of this seed about 3000 plants flourished well, and about 50% carried male blossoms.

With the view of extending this cultivation, Gillavry started in 1886 a nursery in which the fruit of the best trees of the old plantation was planted out. Although the fruit was small, it was used with advantage, as only 35% male trees were produced from it.

The seed rapidly loses its germinating power, and it is therefore necessary to plant it soon; boxes with moist sand are used for transport. A month after sowing out the seed germinates, producing in the first instance a tap-root without ramifications. The young plants must not be taken up until after 18 months, and then only together with a sufficiently large lump of soil. If no favourable site is available for such cultivation, it is advisable to arrange the seed beds in plaited baskets. The plants should grow in a shady place until the trees have developed a sufficient number of branches to shade the soil themselves.

The nutmeg tree is cultivated successfully at altitudes of from 750 to 2000 feet.

The preparation of the crop is very simple. The fruit-pods and mace are removed from the seed, washed in salt water, and dried as quickly as possibly in the sun or in a drying cupboard. The nuts

---

[1]) Revue des cultures coloniales **14** (1904), 342.

from which the pods have been removed are rolled in slaked lime, and packed in cases which have been given a lime wash on the inside; the cases hold 60 kilos nutmegs, and measure $18 \times 18 \times 18$ inches. The mace is packed in cases of $24 \times 24 \times 24$ inches, lined with paper.

This cultivation may be recommended as a by-cultivation, as the nutmeg tree does not require much care; nor much. manure. The tree bears fruit from the seventh year, but the crop is not good until the twelfth year,. and then improves from year to year.

Of the enemies of the nutmeg tree a fungus may be mentioned, which attacks the branches and causes their rapid decay; as the origin of this disease is unknown, it cannot be combated.

Other pests are a beetle which bores holes in the trunk, and a kind of ulcer; trees attacked by the latter, are best felled at once.

Opopanax Oil. The opopanax resin found in commerce, which serves for the production of the opopanax oil used for purposes of perfumery, is, as Holmes[1]) has already pointed out, not identic with genuine opopanax; according to an examination made by A. Baur[2]) in 1895, it originates from a species of *Balsamodendron* belonging to the family of Burseraceæ, probably *Balsamodendron kafal* Kunth. The chemical composition of the essential oil which can be obtained from this resin by steam-distillation, is still unknown. Some results of an examination of this oil recently made, may find a place here.

The oil produced by us had the following constants: $d_{15°}$ 0,895; α_D — $12° 35'$; saponification number 14,5; soluble in 1 vol. 90 per cent. alcohol; does not form a completely clear solution with 8 vol. 90 per cent. alcohol. When the oil was acetylised an increase in the saponification number was observed, which proves the presence of alcoholic constituents.

The bulk of the oil distilled *in vacuo* (3 mm.) from $45°$ to $130°$. The viscid brown distillation residue was heated with phthalic acid anhydride to $100°$, in order to collect any sesquiterpene alcohols which might be present. The reaction-product, worked up in the usual manner, yielded a phthalic ester acid which, on saponification with alcoholic potash, split off an alcohol which was extremely difficult to volatilise with water vapour. The alcohol which was present only in small quantity, distilled *in vacuo* (2 mm.) from $135°$ to $137°$ as a viscid, colourless oil, possessing a peculiar odour reminding of opopanax. Treatment with phenyl isocyanate yielded a crystallising phenyl urethane, whose melting point, however, did not become constant, in spite of repeated recrystallisation. To all appearances the

---

[1]) Pharmaceutical Journal III. **21** (1891), 838.
[2]) Arch. der Pharm. **233** (1895), 209.

5*

alcohol obtained represents a mixture. The fraction of opopanax oil which was distilled off *in vacuo*, boiled at ordinary pressure as a sesquiterpene chiefly from 260° to 270°. This fraction was dissolved in 3 to 4 parts ether, and saturated with hydrochloric acid gas. After evaporating the ether, a large quantity of a crystallised hydrochloride remained behind, which, after repeated recrystallisation from alcohol, melted at 80°. A benzene solution of the hydrochloride proved to be optically inactive.

According to the analysis, the compound appears to have the composition $C_{15}H_{24} \cdot 3\,HCl$. The hydrocarbon of opopanax oil would therefore be a sesquiterpene with 3 double-linkings. The hydrocarbon regenerated from the hydrochloride by boiling with sodium acetate in glacial acetic acid, distilled at 3 mm. from 114° to 115°. At atmospheric pressure it passed over with decomposition from 260° to 285°.

The constants of the regenerated hydrocarbon were:

$$d_{15}\ 0,8708;\quad \alpha_D \pm 0°;\quad n_{D_{26}°}\ 1,48873.$$

If the regenerated hydrocarbon is saturated with hydrochloric acid gas, the original hydrochloride of the melting point 80° is again formed.

Oil of Oregon balsam. In the laboratory of Edward Kremers, Frank Rabak[1]) has submitted to a more detailed examination Oregon balsam on whose origin and composition up to the present the most diverse and partly contradictory opinions prevailed. According to the very meagre literature hitherto published, Oregon balsam is said to have been met with in commerce for the first time in 1874. Dowzard[2]) even considered it a solution of colophonium in oil of turpentine, whose only object was to serve as an adulterant of Canada balsam; such an adulteration, however, would be readily detected.

The balsam examined by Rabak had been supplied by a New York firm, and been obtained from *Pseudotsuga mucronata* Sudworth, as was established by R. H. Denniston by an examination of the branches of trees which, according to authentic information, had been employed for the production of Oregon balsam. Rabak obtained from the balsam, by steam-distillation, 25% of an essential oil, the bulk of which distilled over below 160°. The oil has a pleasant turpentine-like odour. The specific gravity fluctuated in different preparations between 0,822 and 0,873, whilst the difference in the angles of rotation was but slight (α_D — 34° 37' to — 39° 55'). In fractionating the essential oil, 71,8 to 83,4% passed over up to 160°; from this portion relatively pure l-pinene could be obtained by a second fractional distillation. The l-pinene was identified as such, by conversion into pinene nitrosochloride, nitrosopinene, and pinene nitrol benzylamine.

---

[1]) Pharmac. Review **22** (1904), 293.
[2]) Chemist & Druggist **64** (1904), 439.

Orris Oil. Our buyers at Florence report as follows on the market of Florentine orris root: —

According to our previous communications, the quantity at disposal, at the end of February 1904, amounted to . . . . 680 tons
Shipments from the beginning of March to the end of August . 190 „
Consequently on September 1st 1904 still available . . . . . . 490 tons

The total quantity of the shipments in the 12 months ending August 1904 amounted to 820 tons, i. e. 80 tons below the average annual quantity of 900 tons. The new harvest has recently commenced, and will as usual last to October/November. The quality of the crop gathered up to the present leaves much to be desired, — no doubt in consequence of the neglect in the cultivation, and the exceptional drought of the summer of 1904; but later on a better quality may possibly make its appearance, as the harvest in the producing-districts lying at greater altitudes, where rain has repeatedly fallen, only takes place later. With regard to the quantitative result of the harvest, nothing can as yet be stated, nor even an approximate estimate made. The large quantities planted out in 1901 are now ripe for gathering; these must be picked during the present autumn, for reasons mentioned repeatedly. There are further the roots planted in 1902, not quite so large a quantity, which may be gathered either now or in the autumn of 1905, as desired. Now it has often occurred in the past that an untimely advance in the prices exerted an enormous influence not only on the harvest, but also on the new plantings. For example, in the summer of 1902 it was said that only the 3 year old roots would be gathered, which justified the estimate, made at that time, of a crop of 700 to 800 tons. But when towards the end of that harvest the price gradually increased from 38 to 44 marks cif. Hamburg for assorted roots, the growers, by subsequently gathering the 2 year old roots, produced the remarkable total yield of about 1000 tons, and at the same time forgot in many cases their good intentions respecting the planting out of smaller quantities. The foregoing makes it clear that any estimate of the quantity of the harvest just commenced is so far problematical. But if we accept a minimum of 700 tons for the new harvest, we have, after adding the old stocks of 490 tons, in any case an available total quantity of 1190 tons, that is to say 300 tons more than the world's annual consumption. Up to the present a few small sales of the new crop have taken place at the parity of 37 marks cif. Hamburg, for assorted roots; seconds, which are only sold at a later period, might therefore be valued at 32 to 33 marks. It does not look, however, as if for the time being a brisk demand would arise at these figures, as it is reported from many quarters that the previous purchases will last still for a considerable time. During the past twelve months the quotations have not fluctuated greatly, the prices paid were:

	Assorted roots	Seconds
September/December 1903,	35 to 37 marks	31 to 32 marks
January/March 1904,	37 „ 38 „	32 „ 33 „
April/August 1904,	38 „ 37 „	33 „ 32 „

all per 100 kilos cif. Hamburg. In view of the statistical position of this article, an improvement in the prices for the coming season may, unfortunately, hardly be expected.

According to the above reliable report, the exceptionally low prices of orris oil will no doubt remain in force, at least for the next twelve months, provided always that up to the time the roots are gathered no abnormal natural phenomena take place.

We would not miss this opportunity to call attention to the liquid orris oil first introduced into commerce by our firm. This oil meets with a constantly growing demand, and is preferred wherever the myristic acid present in ordinary orris oil has a troublesome effect. This is specially the case in extracts, where possible precipitates on arrival of colder weather may be extremely disagreeable. Our liquid orris oil is exactly ten times as rich as the ordinary oil. It is a product of the very first rank.

Patchouli Oil. The favourable prices paid for the crude material have led to an increased cultivation in Indo-China. In consequence of an arrangement made with a firm at Singapore, we receive regular shipments, and are now able to promptly deliver any quantity of oil. Our correspondents recently informed us that the German steamship-lines are now making great difficulties with the forwarding of patchouli herb, as it is said to have injured other goods by its odour.

We should think that a place could easily be found on board of a steamer, where the herb could be stowed without coming more closely in contact with sensitive goods, among which tea would doubtless rank first.

Simmons[1]) calls attentions to new adulterants of patchouli oil, and emphasises the necessity of determining the ester number, in addition to the usual determination of the physical constants. He has had before him some oils in which the common adulterants cedarwood oil and cubeb oil had been replaced by esters or ester-containing oils. The adulterated oils differed from normal distillates by a lower rotatory power and higher saponification numbers (saponification number 58 and 18,5), which had been produced by an ester of benzoic, or of fatty acid. An attempt made to identify the alcoholic constituent of the esters did not lead to any definite result, although the camphor-like odour pointed to borneol.

In our laboratory similar adulterations of patchouli oil have not yet been observed, although we have already for many years paid attention to such a possible contingency. In our distillates we found the saponification number as 8 to 12,3.

Peppermint Oil, American. In view of the importance of this article, our New York house has endeavoured to make a thorough enquiry respecting this year's harvest, by sending an expert to the respective districts, and has supplied us a detailed report on the results. The principal producing districts are:

----

[1]) Chemist and Druggist **64** (1904), 815.

I. Michigan and Indiana.

In the neighbourhood of Muskegon, Moorland and Revenna, at least 100 out of 400 acres of old peppermint have been destroyed by the severe cold of last winter. About 100 acres have been planted this year with new mint, but they do not promise a good result. Two-year old mint is generally most suitable for distillation, but this year not a yard of such mint is found, as it has been mostly choked by weeds. If 10000 to 12000 lbs. oil are obtained from this district, the result might be called favourable.

Near Fennville and Pearl only about 300 to 400 acres (apart from a few large farms) have been planted with new mint, as compared with 800 to 1000 acres last year, because roots are difficult to obtain. It was reported from various quarters that the largest oil producer had lost 450 acres out of 1100 through the severe cold of last winter.

In the district round McDonald, there are about 300 acres, from which a yield of 4500 lbs. oil is expected, against 6000 lbs. last year.

Round about Decatur, some 1200 to 1500 acres mint are found. A few years ago the whole marshland was cultivated with the herb, but when the oil-prices went down, the majority of the farmers gave up the cultivation. The yield of oil in this district may possibly come to 30000—40000 lbs.

The country to the south of Centreville was formerly also planted largely with mint, but now other plants are cultivated, as mint exhausts the soil to such an extent that it can only be grown for a few years. At present new mint is grown only on 100 acres, and a yield of 3000 to 5000 lbs. oil would be regarded as a good result.

Between White Pigeon and Constantine there are about 300 acres of new mint. One farmer living at Three Rivers owns 50 acres new mint. Of his 200 acres of two-year old mint, 100 have been destroyed by the cold.

The distillation from 5 acres two-year old mint belonging to another farmer yielded in the presence of our informant about 11 lbs. oil. At other times such mint requires no care, and is free from weeds, but this time the weeds, favoured by the weather, have choked the mint before it was able to develop properly.

In the vicinity of Three Rivers, Moore Park, and Parkeville, there are about 400 acres of new mint plantations, and about 800 acres of two- and three-year old mint, but the latter leaves much to be desired.

Near Sturgis, no mint is found; near Burr Oak only 100 acres planted with new mint, and the two- and three-year old mint there does not look promising.

At South Bend, Indiana, one farmer has 247 acres old mint, of which 60 acres have been lost through the cold. As he could not obtain a sufficient number of roots, he was only able to plant 2 acres with new mint.

Between Mishawaka and Osceola the new plantations amount to 150 to 200 acres, and in Mishawaka itself to about 80 acres.

All these statements are based on personal inspection, and our informant is convinced, in view of the condition of the peppermint fields, that the result of the harvest will not exceed two-thirds of that of the previous year, and will consequently amount to about 90000 lbs.

It should still be mentioned that a number of fields from 5 to 15 acres in Cass and Branch Counties are cultivated with mint, but these have not been visited by our informant.

II. Wayne County, New York.

The reports received from time to time that the peppermint cultivation has almost been given up during the last six years, have been found to be correct during a visit to the principal district. Thousands of acres on which formerly only peppermint was grown are now planted with sugar-beet, onions, and celery, as the marshy soil, called "muck land" by the farmers, is particularly suitable for this purpose.

But during a ride through all these lands it was found that the cultivation of peppermint herb, although it had been given up for several years, has again been resumed by the farmers in that part of Wayne County which for many years was considered the most important peppermint oil district of the world.

If the prices of oil remain at their present level, it may be taken for granted that the hundreds of acres on which peppermint is now cultivated will increase in one or two years time to thousands.

In these districts there are farmers who this year (1904) place oil on the market, after not having sold 1 lb. of it during the last five years. An inspection shows that the oil-producers have allowed their distilling-apparatus to go to ruin and become useless. We may mention here that the oil-distillation is usually carried out by one person for all his neighbours. For example, the individual farmer cultivates a few acres; he sends the cut herb some 5 or 10 miles away to the nearest distiller, and pays 25 to 35 cents per lb. for the production of oil. The distiller, usually of a speculative turn of mind, makes an offer to the farmer for the oil. In this manner the former collects considerable quantities, and often proves to be a good source of supply for the oil, or a source of information as to where the oil can be obtained. Places such as Clyde, Lyons, Newark, Palmyra, etc., do not belong to the real producing places; here the oil supplied by the farmers is the object of speculation. The actual producing districts are at least 5 to 6 miles further up country. The average farmer is not a well-informed person, and believes the word of the speculator.

It is certain that the cultivation of peppermint is again on the increase, and the farmers predict that if the prices keep at their present level, Wayne County will again resume its former position, and produce the finest oil in the world.

Not a single acre in Wayne County has been planted with American mint, as the cultivation of this herb has been completely neglected. Only here and there some small spots are planted with it, and not a pound of American peppermint oil is placed on the market under this name.

Black mint yields the highest percentage of oil, and according to the farmers it is the only one found in Wayne County for the production of peppermint oil. With proper care an acre yields 30 to 50 lbs. oil. The farms which are under cultivation have a healthy appearance, and farmers who have attempted the cultivation this year are most hopeful as to the results. As a matter of fact, roots were almost unobtainable, until an enterprising farmer and oil-speculator imported a parcel of roots of black mint from Canada, and sold them at the enormous price of $ 1 per rod (?). About 10 rods of roots are required for planting one acre. The high price of the roots induced many farmers to give up the idea of planting mint, and a large proportion of this year's cultivation has chiefly taken place with the view of obtaining roots for next year. This alone shows that in 1905 an increased cultivation will have to be reckoned upon.

A large number of farmers who have now only planted one half to one acre, will not have the herb distilled, but will keep the roots for next year.

It is not possible to form a correct estimate of the land planted with mint in Wayne County, as the distillation has almost ceased to exist as an industry since six years, and the farms are spread over a large district. It was necessary to ride across many miles of country in order to find actual peppermint farms; on many mint-lands which some years ago where thriving well, other cultivations have since then been taken up.

According to our observations, the total area cultivated with peppermint in Wayne County may possibly amount to about 445 acres.

As the farmers assume that they will obtain on the average 30 to 50 lbs. of oil from one acre, it follows that the new yield of black mint oil in 1904 will be about 14,000 lbs.

It is stated there that the black mint herb is chiefly cultivated on account of its high yield; but the roots do not keep well, and no reliance can be placed on this. The peppermint herb designated by the farmers as "old style" is no longer grown, on account of its low oil-content.

It is found that almost all the old mint has perished during the winter of 1903/04 through cold; unquestionably the production of roots will receive the greatest care, so that the result of next year's crop will turn out considerably higher. The distiller is, as a matter of fact, the principal person, as he not only constantly keeps in touch with distant farmers, but also is a large dealer in oil. As is well known, the bulk of the oil is taken to Lyons or Newark and handed over to speculators in those places.

Formerly peppermint oil distilling-apparatus could be seen from the road everywhere, but with a few exceptions this trade has made room for more remunerative occupations.

This autumn about 10 installations will be at work, whereas six years ago more than 100 were working daily. This alone may explain the scarcity of oil during the last few years.

At the time of the inspection-journey, six months before the commencement of the distillation, the peppermint herb was not yet completely developed, and the result of the harvest as mentioned by the growers could therefore only be an estimate. Occasionally it was difficult to obtain information; the estimates have in every case been carefully checked.

The facts mentioned in the foregoing report would justify high prices for American peppermint oil, if large stocks of the previous harvest are not in existence.

The wild speculation will still be remembered which was intended to drive up the price to $ 6,— per lb., but which came to grief when it became known that much Japanese oil had been used for adulteration, and that the rise had been brought about artificially. It appears that matters will go the same way this year. We do not grudge the farmers good prices, but the clumsy speculation must be combated with every means at disposal.

For the present we recommend the well-tried tactics of avoiding large purchases, and only buying the absolutely necessary quantities. We hope that then after a certain time, the now prevailing excitement will calm down, even if considerably cheaper prices cannot be reckoned upon.

Peppermint Oil, English. According to reports from the English peppermint-districts, the quantity of oil produced from this

year's crop does not amount to more than one half of an average
result. Such a small yield has not been recorded for 20 years. The
bulk of the 1904 distillation has already been sold, and that at prices
which are about 20 % higher than those of 1903.

The quality of this year's Mitcham oil is good.

Under these circumstances the prices were bound to advance, and
they will have to be raised further, as soon as our cheaper stock
remaining over from 1903 has been sold out.

Peppermint Oil, Japanese. The high prices obtained last
year have led to an important increase in the cultivation of peppermint in the principal Japanese districts, to such an extent that the total
production of the first crop in the Bingo-Bitchin district alone is said
to have reached the phenomenal amount of 300000 catties, or about
200000 kilos. The first crop in the Yonezawa districts, which comes
a little later, is estimated at 120000 catties, or about 72000 kilos,
and the total production of Japan during this season, in view of a second
and third crop, may possibly be estimated too low at half a million
catties, or 300000 kilos.

In face of such figures all efforts made by the Japanese to keep
up the prices proved to be unsuccessful, and towards the middle of
August the early signs became apparent of the collapse which occurred
early in September. The price of crystals, which in the autumn of
1903 was about 18/-, fell in the course of a few weeks to 9/-, and
that of dementholised liquid oil from 5/9 to 4/-.

We must wait and see whether a further decline will take place.
In view of the upward tendency in American oil this can hardly be
expected, for there will undoubtedly again be a strong demand for
Japanese oil from America.

The total shipments of peppermint oil, liquid and in crystals,
during the first 6 months of this year, were: —

I. from Yokohama:	crystals	liquid oil
to London	279 cases	228 cases
„ New York	102 „	13 „
„ Paris	$16^{1}/_{2}$ „	$12^{1}/_{2}$ „
„ Hamburg.	$263^{1}/_{2}$ „	$224^{1}/_{2}$ „
„ Bombay	$156^{1}/_{2}$ „	$39^{1}/_{2}$ „
„ Calcutta	17 „	— „
„ Hongkong	16 „	3
„ San Francisco	— „	10 „
„ Samarang	— „	15 „
Total from Yokohama	$850^{1}/_{2}$ cases	$542^{1}/_{2}$ cases

II. from Kobe: crystals   liquid oil

to London	105	cases	20	cases
„ New York	267	„	20	„
„ Hamburg	85	„	179	„
„ Bombay	9		8	
„ Hongkong	61		63	
„ San Francisco	5	„	2	
„ Marseilles	26		13	
„ Singapore	25		—	
„ Cleveland	7		—	
„ Hamilton	2		—	
„ Sydney	—	„	1	„
„ Shanghai	—	„	12	„
Total from Kobe:	592	cases	318	cases.

Summary:

from Yokohama $850^1/_2$ cases crystals $542^1/_2$ cases oil, total 1396 cases
„ Kobe $\quad$ 592 $\quad$ „ $\quad$ „ $\quad$ 318 $\quad$ „ $\quad$ „ $\quad$ „ $\quad$ 910 „
$\overline{}$ $1442^1/_2$ cases crystals $863^1/_2$ cases oil, total 2306 cases

or $1442^1/_2$ cases crystals of 60 lbs. $=$ 86550 lbs.
$863^1/_2$ „ liquid oil „ 60 „ $=$ 51810 „

from Jan. 1st to June 30th 1904 $\quad$ Total crystals and oil 138360 lbs.

Peppermint Oil, Saxon. Whilst all our other cultivations in the neighbourhood of our factory have suffered greatly from the exceptional drought, the crop from our peppermint-plantations has been good. We have obtained a product which in point of quality occupies a high place. Owing to the lack of rain extending over several months, is has not come to a second crop.

Parry and Bennett[1]) have recently detected cedarwood oil as an adulterant of oil of peppermint, in several samples which, according to the results of the examinations, no doubt all come from the same source. The physical constants of the oils in question did not immediately point to adulteration, although the total content of menthol was rather low; but in 70 per cent. alcohol the oils were insoluble. By means of fractional distillation, and comparison with corresponding fractions of pure peppermint oil, the adulterant was determined with a fair amount of certainty as sesquiterpene. The later fractions of the adulterated oil were much less soluble than those of pure oil, and in some the taste of cedarwood oil was clearly perceptible. At the

---

[1]) Chemist and Druggist **64** (1904), 854.

end of their communication Parry and Bennett give a tabulated review of the results of the fractional distillation of pure and adulterated peppermint oils, as obtained by them up to the present.

Petitgrain Oil. The uncertain situation in Paraguay has temporarily diverted the attention of the manufacturers there from this product, and as a consequence nothing whatever has been heard about it for many months. This, as well as the slow shipments, point to a restricted production, and give rise to the surmise that it will be difficult to execute the important contracts in due time.

Pine-needle Oils. The consumption of the various distillates from conifers which are included in the aforesaid designation, has increased during the last few years in such manner that the distilling-plant available in the producing-districts, which is mostly of a primitive character, was not sufficient for coping with the demand, and it was found necessary to erect other more rational plant instead. This plant had unavoidably to be established at a greater distance from the centres of production of the raw material, and this has caused a marked increase in the working-expenses. The oil chiefly affected by the changed conditions was the important pine-needle oil from *Pinus Pumilio*, the price of which had to be raised. The increase was absolutely necessary to ensure continued production on a sufficient scale. We hope that we shall now be able to supply the oil also during the winter without interruption.

We were able to leave the price of the fine, highly odorous oil from *Abies pectinata* unchanged for the present; the same applies to the Swiss distillate from the cones of *Abies pectinata*, although it is occasionally very difficult to obtain the latter. The demand for the fine Siberian pine-needle oil has lately shown a considerable falling off. We can strongly recommend this exceedingly pure and very moderately priced distillate.

Rose Oil, German. The conditions this year for our rose-plantations were as unfavourable as they could possibly be. Owing to lack of moisture, only a very small proportion of blossoms was able to develop. Most of the buds pined away and fell off. The result of the manufacture, in addition to about 20000 kilos rose-water and 1000 kilos rose-pomade, was only about 6 kilos rose oil. The prospects for next year have also been injuriously affected, owing to the fact that the rose-trees have formed no new shoots, and that it was necessary to cut them short; this will naturally have a most unfavourable effect on next year's blossom-harvest, and not for several years will there be again a normal yield.

Rose Oil, Turkish. It is well known that this year's rose-harvest in Bulgaria has taken place under the most favourable conditions, and has given a yield of about 5000 kilos rose oil. If this result is not quite equal to that of the previous year, when about 6 200 kilos were obtained, it yet exceeds that of an average harvest. This result was all the more surprising, because as late as April an advance in the prices was represented as probable, — no doubt with the view of disposing of the stocks of old oil.

Up to the present, about 2 200 kilos have changed hands since the new harvest, and that at prices which lie near 500 marks. Below these prices, so-called first quality oil has not yet been obtainable. Many distillers hold out for decidedly higher prices, and to-day the same comedy is again being enacted as in other years. Our buyer reports that the adulteration of rose oil by producers and merchants is seriously on the increase, and that the difficulty of finding pure otto grows from year to year.

Of the rose-oil factories erected at Karlovo, one, so we are informed, has already gone bankrupt and is now closed. For the two others, the prediction is also unfavourable.

Under pressure of keen competition we supply rose oil at an extremely small profit, which does not exceed the rate of simple interest.

The export of rose oil during last season, i. e. from April 1st 1903 to April 1st 1904, was as follows: —

to France	1834	kilos
„ America	1403	„
„ Germany	1039	„
„ United Kingdom	913	„
„ Turkey	422	„
„ Russia	278	„
„ Austria	55	„
„ Italy	32	„
„ Switzerland	13	„
„ Holland	3	„
„ Belgium	4	„
„ Other countries	6	„
Total	6002	kilos
against	3624	„

during the same period 1902—1903.

On the other hand, the export from January 1st to July 1st 1904 came only to 993 kilos, against 1951 kilos in the same period in 1903.

Under the foregoing conditions an increase in the prices can hardly be expected, for, according to present experience, it seems out of the question that speculators will take up on a large scale an article of which it is difficult to judge the real value.

Rose Oil (artificial) "Schimmel & Co." (German Patent No. 126736). This excellent preparation renders it possible to the perfumer to employ the odour of the rose also for cheaper products. These advantages are universally recognised, and the consumption has already acquired very considerable dimensions. In the case of a possible failure of the harvest in Bulgaria, full advantage would no doubt be taken of this product of German science and industry.

Jeancard and Satie[1]) supply a contribution to the analysis of rose oils. In addition to ordinary rose oil, they have also included other distillates in the sphere of their observations, among which the one obtained from the blossoms after removal of the petals (i. e. from the calyx, stamen and pistil) is particularly interesting. This oil had the following properties: $d_{15°}$ 0,8704; a_D — 41°; solidifying point $+8°$; acid number 6,12; ester number 22,4. The content of stearoptene (which consisted chiefly of a body melting at $+14°$) was $51,13\%$. Of alcohols, the oil, strange to say, only contained citronellol, and that in a quantity of $13,56\%$.

The determinations with normal rose oil were, unfortunately, made in such a manner, that the oil from which the stearoptene had been removed came under examination; this impaired the value of the results obtained, inasmuch as it renders it impossible to compare Jeancard and Satie's results with those of other chemists and of ourselves. For this reason we refrain from quoting here the respective data.

In the case of Bulgarian rose oil, Jeancard and Satie determined the solidifying point at $+19°$ to $+21°$, a fact which on the whole agreed with our observations. But in the opinion of the authors, the solidifying point is by no means a criterion for the stearoptene-content of rose oil, as, for example, a distillate from the tea-rose, which solidified at 23,5°, contained 72 to 74% stearoptene, whereas, according to the other observations, a much higher solidifying point was here to be expected. As stearoptene consists of paraffins of different melting points, the abnormal behaviour observed can be thus explained, that in this case a large quantity of a readily melting paraffin (melting point $+14°$) is present.

For this reason Jeancard and Satie consider it inadvisable to value rose oils simply according to their solidifying points, and they recommend instead the determination of the content of stearoptene and citronellol.

Against this, we would point out that the above statements with regard to the relation between solidifying point and stearoptene-content

---

[1]) Bull. Soc. Chim. III. **31** (1904), 934.

are not quite correct. In the distillation-material which for practical purposes comes exclusively under consideration, viz., the blossoms of *Rosa damascena* Miller, the same mixture of paraffins will always occur, and the solidifying point will here consequently always correspond to the stearoptene-content. This view agrees with present experience. If Jeancard and Satie apply their observations made on other distillates to ordinary rose oil, it follows a priori that such a conclusion is not convincing.

Naturally, the solidification point alone cannot be accepted as a standard of value for rose oil, but in conjunction with the determination of the other properties it will always be useful for judging the value of the oil.

For the purpose of valuing rose oil, F. Hudson-Cox and W. H. Simmons[1]) recommend the determination of the iodine number, which is said to be here very serviceable. It is carried out in the following manner; 0,1 to 0,2 gm. rose oil are diluted with 10 cc. 90 per cent. alcohol, and after adding 25 cc. of Hübl's iodine solution, left standing for 3 hours at the temperature of the room. The process is completed in the well-known manner.

For a series of rose oils which are stated to be pure, and which date from the years 1896 to 1903, Hudson-Cox and Simmons determined as limits of value of the iodine number 187 to 194, whilst the iodine numbers of the adulterants which come under consideration are partly considerably higher.

Whether the determination of the iodine number will be serviceable for testing rose oil, is a question which must depend upon further results. Up to the present this method, which is so useful for fatty oils, had always failed in the case of essential oils, for obvious reasons, and we do not therefore entertain much hope in this instance, the more so as, according to Hudson-Cox' and Simmons' own statements, the method is still capable of improvement in various respects.

The discovery of small admixtures in rose oil is therefore still a matter of difficulty, whilst in the case of coarser adulterations the method of testing as hitherto carried out may be found sufficient. We reserve a final opinion on the proposal made by Hudson-Cox and Simmons, until their prospective further publications have made their appearance.

On the strength of numerous analyses made during the last few years in our laboratories, we are in a position to correct and amplify

----

[1]) The Analyst 29 (1904), 175; Pharmac. Journal 72 (1904), 861.

the data mentioned for Bulgarian rose oil in Gildemeister and Hoffmann's work "The Volatile Oils" (p. 429); we determined the following limits of value for Bulgarian rose oil: $d_{15°}^{30°}$ 0,849 to 0,862, rarely up to 0,863; a_D — 1° 30′ to — 3°; $n_{D25°}$ 1,452 to 1,464; congealing point $+$ 19° to $+$ 23,5°; acid number 0,5 to 3; ester number 8 to 16; total geraniol (geraniol plus citronellol) 66 to 74 %, exceptionally up to 76 %; citronellol 26 to 37 %, generally 30 to 33 %.

The citronellol determination is carried out by formylating; we take for 1 vol. oil 2 vol. 100 per cent. formic acid, and heat the mixture for one hour in a reflux condenser; for the rest the process is the same as in acetylating.

The alcohol $C_{10}H_{18}O$ designated by Hesse and Zeitschel[1] as nerol, is, according to recent examinations by v. Soden and Treff[2]), also present in rose oil in the proportion of 5 to 10 %. The rose nerol isolated by them corresponds in its properties with that obtained from petitgrain oil. The melting point of the diphenyl urethane lay at 52° to 53°. There was also detected in rose oil 1 % eugenol. Among the primary alcohols separated off with phthalic acid anhydride there was found a compound $C_{15}H_{26}O$ in the proportion of about 1 % of the rose oil, which appears to be an aliphatic sesquiterpene alcohol, and which may possibly be identic with farnesol discovered by Messrs. Haarmann & Reimer[3]) in the oils of ambrette-seeds and of cassie blossoms.

Rosemary Oil, Dalmatian. The drought has damaged the Dalmatian rosemary-districts to such an extent, that it was found necessary to make an allowance in the price to the distillers. According to the consular reports in the German "Handelsarchiv" of June last, p. 534, the rosemary-production increases from year to year. In 1903 it amounted to about 17,000 kilos, of which 10,000 were sent to Germany. This year, France may possibly appear as a strong buyer of Dalmatian oil, as in that country the drought has played still greater havoc among the plants than in the Dalmatian islands, and the production of French oil of rosemary will be as insufficient as that of spike oil. We would advise to make trials in good time, to ascertain whether the former oil is not able to replace spike oil in various industries, particularly in the ceramic industry, for there can be no doubt that it will only be possible to execute a small proportion of the orders for spike oil.

[1]) Journ. f. prakt. Chem. II. **66** (1902), 481.
[2]) Berl. Berichte **37** (1904), 1094.
[3]) German Patent No. 149603.

As but little is known of English oil of rosemary, owing to the fact that this oil is not an article of commerce, it was a matter of interest to us to receive recently from Mr. Sawer, of Brighton, an English oil distilled by himself. It had a pleasant, powerful aroma, and showed the following constants: $d_{15°}$ 0,9042; a_D — 2° 49'; a_D of the first 10% — 6° 10'; ester number 9,7; soluble in about 5 and more vol. 80 per cent. alcohol, with very slight turbidity[1]).

English rosemary oil differs from the French and Dalmatian by its lævorotation, an observation which agrees with those made previously[2]). In the case of the two last-named oils, the dextro-rotation is exactly considered a special criterion of purity, and dextro-rotation of the 10% first passing over on fractional distillation is also required. As we have lately observed a lævorotation in the last-named case, even in oils from a reliable source (though only in very few cases), it may be asked whether this requirement can be upheld without qualification, or whether a slight lævorotation of the first 10% of the distillate may be permitted if the specific gravity is sufficiently high?

For the present, we must still insist on our former requirements, and we reserve a definite opinion on this question, until we have had before us a more voluminous authentic material for observation.

Some years ago we published a note on Spanish oil of rosemary[3]), according to which this oil differs from the French and Dalmatian by its higher specific gravity and its more powerful rotation; the ester-content was also higher, as saponification numbers were found up to 37, whilst otherwise they hardly exceed 12. But as we also repeatedly received Spanish oils which were in every respect the same as the French and Dalmatian distillates, we believed that the above-mentioned differences were due to the fact that in those cases the rosemary distillate was not pure. This surmise has now been confirmed. According to information received by us from Spain, there exists in that country, besides the ordinary rosemary oil, which is quite normal in its behaviour, also an oil designated an "rosemary oil courant" which is distilled from rosemary and sage, and which differs in the manner described above from ordinary rosemary oil. An oil of this kind recently examined by us had the following constants: $d_{15°}$ 0,9258; a_D + 14° 35'; a_D of the first 10% + 0° 40';

---

[1]) The oil had been distilled from herb gathered in the autumn (22. Sep. to 30. Oct.). A distillate from the same source, made from hibernated herb cut in April, had approximately the same constants as the one described above: $d_{15°}$ 0,9047; a_D — 2° 28'; ester number 6,52; soluble in about 5 and more vol. 80 per cent. alcohol.

[2]) Sawer, Odorographia, Vol. I, 370.

[3]) Report April **1900**, 37.

acid number 0,9; ester number 35,7; soluble in 1 and more vol. 80 per cent. alcohol.

We shall not fail to devote our constant attention to this subject.

Sandalwood Oil, East Indian. The tendency of the prices has now for some months been upwards, due to the fact that a few manufacturers in the present state of price-cutting were unable to exist. With the unusually strong demand we are fortunate in having sufficiently large stocks of sandalwood to last us until the spring, even when working at full pressure. The demand was occasionally so important that we had to work day and night.

Under such conditions it is useless to send orders at limited prices, as they have no chance of being accepted. The further course of the prices depends on the result of the auctions which take place in India from November 19th to December 19th. It has already been settled that the Government will only offer as much wood for sale as is required for the world's consumption, viz., about 2500 tons, — a procedure which can only be approved of.

According to a note in the German "Handelsarchiv" of March last, Western Australia still continues to carry on a brisk trade in sandalwood which is felled on the banks of the Swan river. In 1903, the value of the export to China and Singapore came to $ 61 771. As already reported on a previous occasion, this variety differs in a marked degree from the more precious East Indian wood, and yields an oil which on account of its resinous odour cannot be used for purposes of perfumery.

_____

The observation made some time ago by Peter[1]), that the oil of sandal capsules is frequently seriously adulterated, has recently again been confirmed by Runge[2]), who reports on this matter in the Pharm. Zeitg. Of two such oils, one was strongly adulterated, as is proved by the following constants: $d_{15°}$ 0,959; α_D $+6° 30'$; santalol content[3]) about 56,5%; insoluble in 10 parts 70 and 80 per cent. alcohol.

This result induces Runge on the one hand to attack the article offered at cheap prices, and on the other, to recommend a thorough test of the sandalwood oil bought in capsules. Where Runge for this purpose quotes among others also Conrady's colour-reaction, we can only emphasise once more that this unscientific method of testing

_____

[1]) Report October **1903**, 63.
[2]) Pharm. Zeitg. **49** (1904), 671.
[3]) Runge speaks of "santalol number"; we presume that he means thereby the content of santalol in per cent. Unfortunately Runge does not state at all whether the results are based upon the formula $C_{15}H_{24}O$, or $C_{15}H_{26}O$. The latter is still used frequently, but we have often pointed out that according to more recent examinations $C_{15}H_{24}O$ is undoubtedly more correct.

is unserviceable for the valuation of sandalwood oil, a fact proved by our numerous experiments, of which some have again been made recently [1]). With regard to Runge's assertion that in normal sandalwood oil the rotation increases with the specific gravity, we must correct this in so far that specific gravity and rotatory power vary here independently of one another; we have been able to establish this by a very voluminous observation-material.

In the Communications from the Biologico-Agricultural Institute, Amani, A. Zimmermann [2]) gives a review of all trees supplying wood designated a sandelwood, of which many are in no way related to *Santalum album* L. We also abstract from this publication some interesting information on the most important one, the East Indian sandalwood; the cultivation of this tree should first of all be taken into account for the German Colony. As the genuine sandal tree which has hitherto only been cultivated in East India and Java, belongs to the root-parasites, this fact should not be lost sight of when laying out plantations. The parasitical mode of life commences already a few months after germination, when the rootlets of *Santalum album* L. drive real haustoria first of all into the roots of grasses, herbs, and small shrubs, subsequently also into those of trees. For this reason the young plants are suitably planted out each along with another young plant in baskets made from the sheaths of bamboo leaves, and are later on best cultivated in mixed stocks. The most advantageous time for the harvest is in the 27th to 30th year, when the trees are felled and the thicker roots dug out. The wood is barked, split up, and assorted, when special regard should be paid to the colour, as the oil content of the wood as a rule is all the larger, the darker the colour is.

Spearmint Oil. In view of the small American production it was necessary to pay liberal prices for this very important article. By doing this, and also by our excellent connections, we were in a position to meet the requirements of our clients.

Spike Oil. The position of this article, which in itself is very critical, will become even more acute in the near future, for the new distillation has given such a small yield of oil, owing to the abnormal drought, that a sufficient supply to meet the world's requirements is out of the question. The sparingly grown plants were nearly dried up when submitted to distillation, and consequently gave only very small yields of oil.

---

[1]) Comp. Report April **1904**, 82.
[2]) Reprint from the "Usambara Post". Communications from the Biologico-Agricultural Institute, Amani. May 21, 1904, No. 25.

We hear from the South of France that the adulteration of spike oil is carried on this year in a particularly impudent manner, chiefly with camphor oil containing safrol, and further with Spanish sage oil, which finds its way almost exclusively to the South of France. The latter has a value of 3 to 4 francs, and is a product which lends itself to all possible adulterations.

It stands to reason that the prices of spike oil had to be raised considerably. In view of the scarcity of this oil, it may be worth while to consider whether, and in how far, Dalmatian rosemary oil can replace spike oil in practice.

Star-anise Oil. The events in the Far East appear to have diverted attention entirely from this Chinese product. Since the spring, the prices have gradually declined from 5/3 to 4/9, and even a transaction of about 500 cases, concluded by us at Hongkong in August, has not led to any advance worth mentioning. This fact points to the existence of large stocks in China.

Speculative stocks bought at high prices are said to be still held in London and Hamburg.

In our opinion the present value of this article is low. A drop below 5/- was at the time declared by the most experienced experts to be impossible.

Turpentine Oil. Worstall[1]) has made experiments with regard to the absorption of iodine by turpentine oil, and he finds in the determination of the iodine number a serviceable means of establishing adulterations of oil of turpentine, if the work is carried out under exactly identical conditions.

Worstall lets 0,1 gm. oil with 40 cc. Hübl's iodine solution stand overnight in a bottle provided with a glass stopper, and then titrates back the excess of iodine. The average iodine number of 55 pure turpentine oils was found at 384, whilst in theory 373 was to be expected on converting $C_{10}H_{16}$ into $C_{10}H_{16}I_4$.

Further experiments showed that the absorption of iodine according to the above directions is accomplished in 4 to 6 hours, and that with a large excess and prolonged action of the iodine solution, à further absorption takes place through secondary reactions.

The iodine number of resin spirit was determined at 185, of resin oil at 97, of refined wood turpentine at 212, and of water-white at 328, whilst kerosene and naphtha, as might be expected, did not absorb iodine.

Slight admixtures of petroleum distillates will therefore show themselves already by a lowering of the iodine number; in any case,

---

[1]) Journ. Soc. chem. Industry **23** (1904), 302. Chem. Centralbl. **1904**, I. 1676.

a turpentine oil which under the above conditions gives an iodine number below 370, must always (according to Worstall) be looked upon with suspicion.

The same subject has occupied the attention of Harvey[1]) who assumed that it would be possible to detect adulterations of lemon oil with oil of turpentine by determining the iodine number. In theory, limonene should absorb 4 atoms halogen, pinene, however, only 2 atoms; the calculated iodine numbers are 372 and 186 respectively. In support of his view, Harvey quotes the iodine numbers 166, 198 and 221 of three American oils of turpentine; the first 10% of the distillates of six samples of lemon oil, however, gave iodine numbers between 334 and 349.

More detailed experiments made on account of the differences existing between the iodine numbers of turpentine oil, led to the result that the quantity of halogen absorbed by the turpentine oil depends not only on the period of action and the excess the Wijs' solution employed, but also on the character of the halogen present in excess in that solution (whether I or Cl in the mixture with I Cl), and that it is better to add the potassium iodide before diluting.

In comparing the results obtained by these two authors, it is clearly seen that in determining the iodine number of turpentine oil, comparable values will only be obtained if the work is carried on under the same conditions.

Edward Kremers[2]) reports on two essential oils which had been obtained by dry distillation of pine roots rich in resin, and which therefore may be designated as pine tar oils. Both samples originated presumably from Georgia, and on the whole resembled in their properties ordinary pine tar oil; they were only somewhat lighter; d 0,856 and 0,860 respectively, after rectification (distillation of the oil shaken with 5 per cent. soda liquor, with water vapour) 0,854; $a_D + 13° 40'$. The oils boiled chiefly between 154° and 180°, and appear to consist principally of pinene, and further of dipentene.

Turpentine oil, which hitherto had been employed as an antidote only in cases of phosphorus poisoning, has recently been used by Allen[3]) for horses with good result as an antidote against carbolic acid; according to the same author, turpentine oil had also an excellent counteracting effect in the case of a man who had taken carbolic acid by mistake.

---

[1]) Journ. Soc. chem. Industry **23** (1904), 413.
[2]) Pharmaceut. Review **22** (1904), 150.
[3]) Apotheker Zeitung **19** (1904), 447.

Oil of Umbellularia californica. Fr. B. Power and Fr. H. Lees[1]) report on an detailed examination of the essential oil of the Californian laurel *Umbellularia californica* Nuttall. This tree which is distributed in the valleys of Oregon and California, contains in the leaves an oil with a pungent odour causing the flow of tears, which had already been examined earlier[2]) by Heany[3]) and Stillman[4]). The latter observed that the body with the pungent odour is chiefly present in the fractions boiling between 215° and 216°, and gave to the compound the name umbellol. The analysis showed the composition $C_8 H_{12} O$.

The quantity of oil examined by Power and Lees amounted to 1200 kilo; $d_{16°}$ 0,9483; α_D — 22°; the oil was completely soluble in 1,5 parts 70 per cent. alcohol, did not react with bisulphite liquor, and contained neither nitrogen nor sulphur compounds. Apart from traces of formic acid and higher fatty acids, the oil contained of compounds soluble in soda liquor only eugenol. Of the latter 1,7 % were obtained by shaking with liquor. The benzoyl compound of eugenol produced melted at 69° to 70°.

Of well-known bodies there were further detected in the oil boiling from 150° to 250°, l-pinene 6 %, cineol 20 %, eugenol methyl ether 10 %, and small quantities of safrol. Of special interest was the fraction 217° to 222°, which represents 60 % of the oil, and which contained the constituent with the pungent odour. This fraction consists chiefly of a ketone $C_{10} H_{14} O$, which Stillman had before him in the impure state. This ketone, on which F. H. Lees[5]) gives further information in a separate publication, is called by the authors umbellulone. It does not react with bisulphite liquor, absorbs 2 atoms bromine, and immediately discolours potassium permanganate solution. The constants of the umbellulone fraction were as follows: boiling point 218°, at 752 mm.; $d_{15,5°}$ 0,9614; α_D — 36° 33'. With semicarbazide it combines to a compound $C_{10} H_{15} (:N \cdot NH \cdot CO \cdot NH_2) CH_4 ON_3$ of the melting point 217°, which is split up by boiling with dilute mineral acids, with separation of umbellulone. The pure umbellulone thus obtained distils from 219 to 220° at 749 mm. as a colourless oil which at first has a pleasant mint-like odour, but on stronger smelling causes the flow of tears.

With hydroxylamine the compound $C_{10} H_{15} (:N \cdot OH) NH \cdot OH$ was obtained as a faint green amorphous mass.

[1]) Journ. Chem. Soc. **85** (1904), 629.
[2]) Report October **1887**, 47.
[3]) Americ. Journ. Pharm. **47** (1875), 105. — Pharmaceutical Journal III. **5** (1875), 791.
[4]) Berl. Berichte **13** (1880), 630.
[5]) Journ. chem. Soc. **85** (1904), 639.

Umbellulone reacts with semicarbazide and hydroxylamine in precisely analogous manner as certain unsaturated ketones with double-linking in the a, β-position towards the carbonyl-group do according to the researches by H a r r i e s, T i e m a n n, R u p e and S c h l o c h o f f. Its behaviour towards bromine shows further, that o n l y o n e double-linking occurs in the molecule; for this reason L e e s considers umbellulone an a-β-unsaturated c y c l i c k e t o n e with two closed rings. U m b e l l u l o n e b r o m i d e, when heated at reduced pressure, is split up into an unsaturated b r o m o k e t o n e $C_{10}H_{13}OBr$ (boiling point $140°$ to $145°$ at 20 mm.) and a dibromo-dihydro umbellulone $C_{10}H_{14}OBr_2$ (melting point $119°$ to $119,5°$). The former can be reduced with zinc dust and acetic acid into the saturated ketone $C_{10}H_{16}O$ (boiling point $214°$ to $217°$). But when dibromo-dihydro umbellulone is treated in this manner, only one bromine atom is eliminated, with formation of bromo-dihydro umbellulone (melting point $58°$ to $59°$). This, however, can be further reduced with sodium in alcoholic solution to tetrahydro umbellulol (boiling point $207°$ to $208°$ at 760 mm.; $d_{15°} 0,9071$; $[a]_D$ —$6,6°$). Both dibromo-dihydro umbellulone and bromo-dihydro umbellulone behave as saturated compounds.

On oxidation with cold 3 per cent. potassium permanganate solution, umbellone readily passes over into a lactone $C_9H_{12}O_2$ of the boiling point $217°$ to $221°$. Tetrahydro umbellulol is a colourless liquid possessing a c a m p h o r- and b o r n e o l-like odour.

Waterfennel Oil. During the examination of an oil of water-fennel freed from phellandrene, we observed that the oil imparted a deep red colour to fuchsin sulphurous acid. When shaken with sodium bisulphite solution a crystalline bisulphite double-compound soon separated off from which, after washing with alcohol and ether, an aldehyde was liberated by means of soda. In order to remove all impurities not reacting with sodium bisulphite, the same purification-process was repeated, and the aldehyde then distilled, first with water-vapour and subsequently *in vacuo*. Apart from the first runnings which contained only a few drops, it had the constant boiling point $89°$ (5 mm. pressure) and the following properties: $d_{15°} 0,9445$; a_D — $36° 30'$; $n_D 1,4911$. Its odour reminds strongly of cuminic aldehyde. On closer examination the aldehyde was found to be an isomeride of citral $C_{10}H_{16}O$, to which we would give the name "p h e l l a n d r a l". In order to identify it, we have first of all produced some derivatives and analysed these, and subsequently made some oxidation tests so as to obtain an insight into its constitution.

The semicarbazone was produced in the usual manner in acetic solution. It crystallises from alcohol (in which it is rather difficult to dissolve), in the form of small crystals of the melting point $202°$ to $204°$.

0,1016 gm. of the substance: 0,2362 gm. CO_2; 0,0836 gm. H_2O.

	Found	Calculated for $C_{11}H_{19}ON_3$
C	63,4%	63,1%
H	9,14%	9,1%

Attempts made to regenerate the aldehyde from the semicarbazone by means of sulphuric acid, did not succeed, as the aldehyde on being set free resinified for the greater part.

The oxime is formed very readily, and quantitatively, when the aldehyde in alcoholic solution is mixed with hydroxylamine hydrochloride and potash liquor. After cooling, the oxime, when water is added, is precipitated as a crystalline powder, and is obtained by crystallisation from alcohol in the form of large brilliant tablets of the melting point 87° to 88°.

0,1364 gm. of the substance: 0,3599 gm. CO_2; 0,1254 gm. H_2O.

	Found	Calculated for $C_{10}H_{16}NOH$
C	71,96%	71,92%
H	10,15%	10,2%

The phenylhydrazone is not very suitable for identification, as it resinifies extremely readily, especially if not quite pure. It crystallises from alcohol in long white needles of the melting point 122° to 123°.

Oxidation of phellandral.

As the aldehyde readily oxidises on exposure to the air, about 3 gm. were left for 3 or 4 days on a rather large watch glass. After the lapse of this period the bulk had solidified in crystalline form. In order to remove any non-acid constituents, the reaction-mass was dissolved in soda solution, well extracted with ether, and the organic acid then precipitated with sulphuric acid. It is very little soluble in water, readily in hot alcohol, from which it is obtained in the form of small needles of the melting point 144° to 145°.

0,1389 gm. of the substance: 0,3643 gm. CO_2; 0,1146 gm. H_2O.

	Found	Calculated for $C_{10}H_{16}O_2$
C	71,53%	71,4%
H	9,17%	9,5%

The silver salt of the acid forms flakes insoluble in water.

0,2664 gm. of the substance: 0,1062 gm. Ag.

	Found	Calculated for $C_{10}H_{15}O_2Ag$
Ag	39,86%	39,36%

The same acid was also obtained by oxidation with moist silver oxide.

Whereas the oxidation of phellandral into an acid with an equal number of carbon atoms only proves the aldehyde character, oxidation tests with permanganate gave some explanation as to its constitution. 5 gm. aldehyde were treated with a 3 per cent. permanganate solution at ordinary temperature, until no further discolouration took place. After driving off with water-vapour any portions which might not yet have become oxidised, the liquid filtered off from the manganese slime was evaporated to about 200 cc., and then acidified with sulphuric acid. A slight turbidity now occurred, and after a few hours about 0,1 gm. crystals separated off, from which the liquid was filtered off. As the bulk of the oxidation product must have been dissolved in water, the liquid was repeatedly extracted with ether, and by these means an acid was abstracted which solidified soon after evaporation of the ether. After recrystallisation from water or benzene-petroleum ether, it melted at 70° to 72°. The acid readily dissolves in the usual organic solvents, and is only insoluble in light petroleum.

9,0966 gm. of the substance: 0,2030 gm. CO_2; 0,0732 gm. H_2O.

	Found	Calculated for $C_9H_{16}O_4$
C	57,3%	57,4%
H	8,4%	8,5%

The silver-salt of the acid dissolves with difficulty in water.

0,3736 gm. of the substance: 0,3658 gm. CO_2; 0,1154 gm. H_2O; 0,2012 gm. Ag.

	Found	Calculated for $C_9H_{14}O_4Ag_2$
C	26,7%	26,7%
H	3,52%	3,43%
Ag	53,7%	53,8%.

The principal product of the oxidation with permanganate is therefore a dibasic acid $C_9H_{16}O_4$. The above mentioned crystalline acid which is difficultly soluble in water, melted, after being recrystallised twice from dilute alcohol, at 115° to 116°. Owing to scarcity of material, no examination could be made with regard to the character of this acid.

The formation of an acid $C_9H_{16}O_4$ from the aldehyde $C_{10}H_{16}O$ is explained most simply by accepting a tetrahydrated benzene-nucleus, in which the double-linking is situated in the β, γ-position towards the aldehyde-group. The oxidation gives no information with regard to the side-chain, but the optical activity of the aldehyde requires an asymmetric carbon atom. In view of the odour which reminds in a pronounced manner of cuminic aldehyde, it may not be improbable that the new aldehyde is a tetrahydrated cuminic aldehyde (p-isopropyl-tetrahydrobenzaldehyde).

Assuming that this view is correct, the acid $C_9H_{16}O_4$, on dry distillation of the calcium salt, should be converted into isopropyl-1-pentanone-3, and the latter, on oxidation with chromic acid, yield a- and β-isopropylglutaric acids. The properties of the former are well known.

In order to carry out this experiment, 6 gm. acid were neutralised with lime, well dried, and after mixing it with an equal weight of soda lime, distilled from a glass retort. The distillate was shaken up with semicarbazide solution, and the resulting semicarbazone repeatedly recrystallised from alcohol; melting point 187°.

0,1034 gm. of the substance: 0,2244 gm. CO_2; 0,0842 gm. H_2O.

	Found	Calculated for $C_9H_{17}N_3O$
C	59,18%	59,0%
H	9,05%	9,3%

The ketone regenerated from the semicarbazone by boiling with 10 % sulphuric acid, boils at 190° to 193°, and possesses an extremely powerful odour reminding of menthone and thujone.

The oxidation of the ketone $C_8H_{14}O$ into the two isomeric glutaric acids $C_8H_{14}O_4$ was accomplished with Beckmann's mixture. The acids abstracted from the oxidation liquid with ether, could unfortunately not be obtained up to the present in crystalline form and identified by these means.

The portions of waterfennel oil which did not react with sodium bisulphite, were first saponified and then submitted to fractional distillation *in vacuo*. The bulk passed over from 60° to 145° (5 to 6 mm. pressure). In the residue there remained a considerable quantity of resin. From the low-boiling portions it was possible to isolate, by repeated careful fractionating, an alcohol of the following physical constants: boiling point 197° to 198°, $d_{15°}$ 0,858; $a_D - 7° 10'$; $n_{D20°}$ 1,44991. The alcohol possesses in a high degree the characteristic odour of waterfennel oil, and appears to be the principal bearer of that odour. With carbanile we obtained a phenylurethane, which after recrystallisation from alcohol melted at 42° to 43°.

I. 0,1002 gm. of the substance: 0,2717 gm. CO_2; 0,0838 gm. H_2O.
II. 0,1208 gm. „ „ „ 0,3286 gm. CO_2; 0,1032 gm. H_2O.

	Found		Calculated for $C_{17}H_{25}O_2N$;	for $C_{17}H_{23}O_2N$
	I.	II.		
C	73,8	74,2%	74,14	74,7%
H	9,3	9,49%	9,04	8,4%

The alcohol accordingly appears to have the composition $C_{10}H_{20}O$, and not to be known up to the present. On oxidising it, we were unable to isolate either aldehydes or ketones. In agreement with its presence in oil of *Phellandrium aquaticum*, we propose to give it the name "androl".

The examination of the higher-boiling portions is not yet concluded.

We would still mention that we were able to isolate from the portions boiling about 230°, by means of phthalic acid anhydride, an alcohol with a rose-like odour, which yielded a diphenylurethane of the melting point 87° to 90° (after one single recrystallisation from alcohol). The quantity was too small to enable us to make a further examination.

Wintergreen Oil. The trade in this American product is one of the specialities of our New York house, which obtains its supplies from sources having proved trustworthy by many years' experience. Naturally, these producers have not been able to escape the influence of the artificial oil, and they work with a very small profit.

The prices of the artificial oil have had to follow the further decline in the values of salicylic acid and methyl alcohol.

Wormwood Oil. After a prolonged interruption we have again obtained supplies of genuine American oil, from which we conclude that the high prices have promoted the cultivation of wormwood in America.

In the South of France the cultivation of wormwood has increased considerably, as offers are becoming more frequent. In point of quality the supplies leave nothing to be desired.

Ylang Ylang Oil "Sartorius". Our friends at Manila have been able to reduce the price of their oil slightly, a step which points to a return of a more orderly state of affairs in the purchase of the blossoms. Recent reports have not been received, but arrangements have been made which ensure sufficient stocks by the commencement of the principal period of consumption.

According to the reports from the German Consulate at Manila, the value of the exported ylang ylang oil was:

in 1902 $ 70 553,—
„ 1903 „ 103 989,—

The last-named figure probably equals a total quantity of about 1 200 kilos. Exceptionally fine qualities, such as the Sartorius brand, which stands at the head, can always be sold quickly and without trouble, but medium qualities remain difficult to dispose of, and are frequently replaced by fine cananga oil.

Russian Pharmacopœia.

The new Edition of the Russian Pharmacopœia published in the year 1902 affords us an occasion for discussing in the same manner as with the other Pharmacopœias, the articles relating to essential oils, in order to correct, or supplement (as the case may be), various statements respecting the oils. This appears to us all the more desirable, as a perusal of the work gives the impression that the Russian Pharmacopœia Committee has not paid sufficient attention to the recent literature on the subject of essential oils.

In order to discuss the details, we reproduce below the requirements specified for the oils. We wish to point out that the strength of alcohol is given in per cent. by volume.

Anise Oil *(Ol. anisi).* Colourless or slightly yellowish; $d_{15°}$ 0,980 to 0,990[1]); a_D not to the right; congeals below $+10°$ to a crystalline mass, which melts again at $+17°$ to $+21°$[2]); soluble in 2 to 3 parts 90 per cent. alcohol, the solution should be neutral, and not be coloured violet by ferric chloride; for medicinal purposes anise oil possessing higher congealing and melting points ($+15°$ to $21°$, and $22°$) is permitted[3]).

[1]) As an anethol separation sometimes occurs spontaneously already at 15°, it is advisable to determine the specific gravity at 20°; the limits of value remain the same also under these conditions.

[2]) This is not sufficiently precise. The congealing point of good anise oil lies above 17°. For the purpose of determination, the oil is cooled down below its solidifying point, say about $+13°$, and if necessary the solidification is promoted by inoculation with a small crystal of anethol.

[3]) This clearly means that for medicinal purposes anethol may also be employed (congealing point 21° to 22°).

Bergamot Oil *(Ol. bergamottae).* Yellowish or greenish; $d_{15°}$ 0,880 to 0,886; a_D to the right; residue on evaporation 4 to 6%[1]); must form a clear solution with a feeble acid reaction when mixed with $1/_2$ volume 90 per cent. alcohol.

[1]) The residue on evaporation in the case of pure oils should be between 5 and 6%; a residue of less than 5% would point to adulteration with oils of turpentine, sweet orange, etc.

Cade Oil *(Ol. cadinum).* A thick, dark brown liquid, with a peculiar empyreumatic odour; lighter than water[1]); dissolves with difficulty in water[2]), more readily in 90 per cent. alcohol; is readily soluble in ether[3]), chloroform, carbon disulphide, and amyl alcohol.

[1]) Occasionally cade oil is also slightly heavier than water.

[2]) The oil ist almost insoluble in water; the latter acquires an acid reaction when shaken with the oil.

[3]) From the solution, which is at first clear, flakes are generally separated off after a short time.

Cajuput Oil *(Ol. cajuputi).* Bright green or yellowish with neutral reaction; $d_{15°}$ 0,915 to 0,930; soluble in alcohol and ether, insoluble in carbon disulphide[1]); when 20 drops cajuput oil are shaken with 5 cc. water and 10 drops nitric acid, the mixture must not be coloured blue after an excess of ammonia has been added.

[1]) All essential oils are soluble in carbon disulphide, and any cloudiness which thereby occurs must be attributed to a slight content of water due to the process of manufacture of the oil.

Clove Oil *(Ol. caryophyllorum)*. Brownish[1]); $d_{15°}$ 1,045 to 1,070; boiling point 250° to 260°; soluble in 2 volumes 70 per cent. alcohol; when 5 drops oil are shaken with 10 cc. lime water, a flaky precipitate should be formed, which deposits itself on the walls of the vessel; identity reaction and test for phenol with ferric chloride in the usual manner.

[1]) Clove oil has a yellowish colour, but becomes darker with increasing age.

Fennel Oil *(Ol. foeniculi)*. Colourless or yellowish; $d_{15°}$ 0,965 to 0,985; a_D to the right; at $+3°$ it solidifies to a crystalline mass[1]); soluble in alcohol; the solution must be neutral and must not be changed by ferric chloride.

[1]) Solidification does not occur spontaneously at this temperature, but must be brought about by inoculation with a small crystal of anethol.

Lavender Oil *(Ol. lavandulae)*. Colourless or yellowish; $d_{15°}$ 0,885 to 0,895; a_D to the left; saponification number not below $84 = 29,4 \%$ linalyl acetate; soluble in 2 parts 90 per cent., and in 3 parts 70 per cent. alcohol; the oil becomes viscid when exposed to the atmosphere, and then acquires an acid reaction.

Lemon Oil *(Ol. citri)*. Clear or slightly turbid; bright yellow; $d_{15°}$ 0,855 to 0,865[1]).

[1]) The range of this constant is so wide that the majority of adulterated oils will probably also answer these requirements; in the case of pure oil, the specific gravity ranges between 0,857 and 0,861 (15°).

Mace Oil *(Ol. macidis)*. A somewhat viscid[1]), clear, yellowish liquid; $d_{15°}$ 0,890 to 0,930; a_D to the right; soluble in 3 parts 90 per cent. alcohol, in an equal volume carbon disulphide; in the last-named case cloudiness occurs when more solvent is added.

[1]) This statement refers to old, resinified oil; the fresh distillate is a mobile liquid, and colourless or yellowish.

Menthol. The requirements correspond to those of the German Pharmacopœia.

Mustard Oil *(Ol. sinapis aethereum)*. Colourless or yellowish; $d_{15°}$ 1,018 to 1,025; $a_D \pm 0°$; boiling point 148° to 152°; soluble in every proportion in 90 per cent. alcohol; the further tests also correspond to those of the German Pharmacopœia.

Peppermint Oil *(Ol. menthae piperitae)*. Colourless or yellowish, in course of time becoming thicker and darker; $d_{15°}$ 0,900 to 0,910; a_D to the left; soluble in 4 to 5 parts 70 per cent. alcohol; when equal volumes water and oil are shaken, and the mixture left standing, two transparent layers should be formed.

Pine Needle Oil (from *Pinus silvestris*) (*Ol. pini foliorum ; ol. pini silvestris*)[1]. Colourless or bright yellow; d_{15° 0,870 to 0,890; soluble in 7 parts 95 per cent. alcohol, less readily in 70 per cent. alcohol.

[1] As the oil distilled from the needles of *Pinus silvestris* L. has a less pleasant odour, and cannot be obtained in commerce, and as moreover the name *Ol. pini silvestris* is an old but incorrect designation of oil from cones of *Abies alba*, it would be more correct to require *Ol. templini*, whose specific gravity lies between 0,850 and 0,870.

Rose Oil (*Ol. rosae*). Colourless[1]), or faintly yellowish; d_{15° 0,87 to 0,89[2]); readily soluble in ether and chloroform, less readily in 90 per cent. alcohol[3]); at $+18^\circ$ to 21° separation of needle-shaped crystals; at $+5^\circ$ solidification to a crystalline mass; when 1 part oil of 0° is dissolved in 5 parts chloroform, and 20 parts 90 per cent. alcohol are added, brilliant crystals separate out after some time; the liquid separated off from the crystals must not impart a red colour to blue litmus paper[4]).

[1] Rose oil is never colourless, but always yellowish.

[2] As the oil, owing to separation of paraffin, is usually already solidified at $+15^\circ$, it is advisable to determine the specific gravity at a higher temperature: $d\frac{30}{15^\circ}$ 0,849 to 0,863.

[3] In consequence of its paraffin-content, rose oil does not, as a matter of fact, form a clear solution with 90 per cent. alcohol.

[4] This test is untenable, as rose oil is slightly acid; acid number 0,5 to 3,0.

Rosemary Oil (*Ol. rosmarini*). Colourless or yellowish; d_{15° 0,90 to 0,92; a_D to the left[1]); soluble in every proportion in 90 per cent. alcohol[2]).

[1] Pure rosemary oil should always be dextrogyre; laevorotation would point to adulteration with French oil of turpentine. Compare the present Report p. 82.

[2] Initial turbidity must be attributed to the water which is always present in the oil.

Sandalwood Oil (*Ol. santali*). Faintly yellowish; d_{15° 0,970 to 0,980[1]); boiling point 300°[2]); soluble in 5 parts 70 per cent. alcohol.

[1] The specific gravity lies between 0,975 and 0,985 (15°).

[2] An actual boiling point in the case of an oil which consists of several components, is out of the question.

Spearmint Oil (*Ol. menthae crispae*). Colourless or yellowish, becoming thicker and darker on exposure to the atmosphere; d_{15° 0,900 to 0,940; readily soluble in 90 per cent. alcohol[1]).

[1] The dilute solution has an opalescent turbidity.

Thyme Oil (*Ol. thymi*). Colourless or yellowish[1]); d_{15° 0,90 to 0,91[2]); soluble in every proportion in 90 per cent. alcohol; phenol-content at least 20%.

[1] Even rectified oils frequently acquire again rapidly the red-brown colour of the crude oil.

[2] As the oils have occasionally a higher specific gravity than 0,91, it would be more correct to require an oil whose specific gravity does not lie·below 0,90.

Thymol. The tests correspond to those of the German Pharmacopœia.

Turpentine Oil, crude *(Ol. terebinthinae crudum).* Colourless or yellowish; $d_{15°}$ 0,860 to 0,875; boiling point 150° to 162°; soluble in 8 to 10 parts 80 per cent. alcohol[1]); when exposed to the atmosphere the oil becomes viscid, and acquires a yellow colour and at the same time an acid reaction; a mixture of equal volumes oil of turpentine and aniline should remain clear.

[1]) Oil of turpentine is much more difficult to dissolve in 80 per cent. alcohol; the above data probably refer to 90 per cent. alcohol, of which usually 6 volumes already suffice; only old resinified oils are more readily soluble.

Turpentine Oil, pure *(Ol. terebinthinae rectificatum).* Obtained by distillation of French oil of turpentine; colourless; $d_{15°}$ 0,855 to 0,860[1]); boiling point 160°[2]); soluble in 12 parts 90 per cent. alcohol; 1 drop oil must not impart a red colour to moistened litmus paper; 5 grams oil must evaporate without resinous residue.

[1]) More correct is : 0,855 to 0,870.
[2]) Pure oil of turpentine passes chiefly over between 155° and 160°.

Novelties.

We are indebted to the kindness of a Japanese firm for a consignment of an oil which is obtained from the leaves and young twigs of the Japanese cinnamon or cassia bark tree, *Cinnamomum Loureirii* Nees, and which is called in Japan **Oil of Nikkei.** The nikkei tree is found in the hottest parts of Japan, viz., in the provinces Tosa and Kii. The oil, obtained in a yield of about 0,2%, has a bright yellow colour and possesses a pleasant odour, reminding of citral and Ceylon cinnamon oil. The other properties of the oil are as follows: $d_{15°}$ 0,9005; a_D — 8°45′; acid number 3,01; ester number 18,6; soluble in 2 to 2,5 and more vol. 70 per cent. alcohol, with opalescence; forms a clear solution with an equal volume and more 80 per cent. alcohol. The oil contains 27% aldehydes, which consist chiefly of citral (melting point of the a-citryl-β-naphthocinchoninic acid 199°). On fractionating the non-aldehydic portions, cineol and linalool ($d_{15°}$ 0,8724; a_D — 27°; $n_{D20°}$ 1,46387) were found; of the last-named compound probably at least 40% is present in the oil.

Contrary to a distillate obtained previously by Shimoyama[1]) from the root bark of the same tree, which was also considerably heavier than the oil here discussed, the one examined by us contained no cinnamic aldehyde.

[1]) Comp. Gildemeister and Hoffmann, "The Volatile Oils", p. 391.

We have also to mention two of our own distillates, made from plants cultivated by ourselves. Unfortunately the quantity of oil obtained in each case was so small, that we were compelled to restrict ourselves to a determination of the properties mentioned below.

Oil of Tanacetum boreale Fisch. The half-dried herb yielded $0,12^0/_0$ oil having a yellowish colour and a powerful thujone odour. d_{15° $0,9218$; $a_D + 48^\circ 25'$; forming a cloudy solution with about 8 vol. 70 per cent. alcohol, with abundant separation of paraffin. The foregoing shows that the oil from this plant which is indigenous to Siberia, fairly closely resembles common oil of tansy in its properties.

Oil from Monarda didyma L. The oil, obtained in a yield of $0,04^0/_0$ from the half-dried herb, had a golden-yellow colour and possessed a pleasant aromatic odour somewhat like ambergris. It had the following constants: d_{15° $0,8786$; $a_D - 24^\circ 36'$; soluble in 1,5 to 2 volumes 70 per cent. alcohol; when further diluted it rapidly became strongly turbid on account of separation of paraffin.

These properties differ slightly from a distillate of *Monarda didyma* which was described about a year ago by Brandel[1]), but this may possibly be due less to climatic conditions, than to a difference in the method of distillation.

Notes on recent scientific work concerning terpenes and terpene derivatives.

The attempts made by scientists to discover by means of synthesis the constitution of compounds having a complicated structure, has recently led to the confirmation of Bredt's camphor formula, and thereby to the solution of a long-pending question[2]). W. H. Perkin jun.[3]) now has followed in the same steps, by producing synthetically the two terpineols of the melting points 35° and 32°. In both cases he started from the δ-ketohexahydrobenzoic acid ester which he had previously obtained by a series of reactions from β-iodopropionic acid ester and sodium cyanacetic ester, that is to say, two bodies produced by synthesis. The above-mentioned ester, when treated with magnesium and methyl iodide according to Grignard, was converted into p-oxyhexahydro-p-toluylic acid ester, whose hydroxyl-group was substituted by halogen. On boiling this halogenised ester with pyridine or weak alkalis, there resulted an unsaturated ester,

[1]) Comp. Report October **1903**, 48.
[2]) Berl. Berichte **36** (1903), 4332. —Report April **1904**, 98.
[3]) Journ. Chem. Soc. **85** (1904), 654.

which on repeated treatment with magnesium and methyl iodide yielded terpineol of the melting point 35°. The complete identity with the natural product was proved by the corresponding constants, the elder-like odour, and the most important derivatives (nitrosochloride, phenyl urethane, terpin hydrate, dipentene, and its derivatives). The terpineol [1]) melting at 32° which was discovered in our laboratories was obtained by Perkin from a by-product (which could not be saponified) formed in the production of the hydrated oxytoluylic acid ester, viz., oxyisopropyl-p-ketohexamethylene, from which by loss of water iso-propenyl-p-ketohexamethylene was formed. Further treatment with magnesium methyl iodide finally yielded the isomeride melting at 32°, which by the formation of the phenyl urethane and the nitrosochloride was identified with the terpineol isolated by us, although the small quantity did not allow of recrystallisation and purification of the alcohol. The following graphic formulae may show the process of the synthesis:

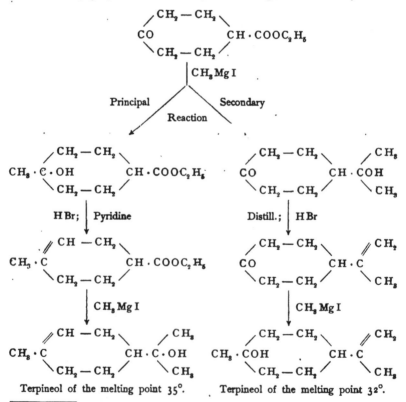

Terpineol of the melting point 35°. Terpineol of the melting point 32°.

---

[1]) Report April 1901, 75. — Berl. Berichte 35 (1902), 2147.

In his 66th Treatise on terpenes and essential oils O. Wallach[1]) reports on the addition-products of nitrous acid and of nitrosyl chloride to unsaturated compounds. After a historical introduction, the author discusses the formula of those yellow compounds formed by the action of alcoholic potash or of acetyl chloride on the nitrites of the compounds containing the group $RCH = CHR$. Here he explains the grounds which induced him to abandon his earlier formula

$$R - CH - CR$$
$$\diagdown O \diagup |$$
$$NO$$

and to accept the formula drawn up by Angeli and lately by Wieland

$$R - CH = CR$$
$$|$$
$$NO_2$$

from which, for the nitrites and the nitroso chlorides, the formulae

$$\begin{bmatrix} R - CH - CHR \\ | \quad\quad | \\ NO \quad NO_2 \end{bmatrix}_2 \quad \text{and} \quad \begin{bmatrix} R - CH - CHR \\ | \quad\quad | \\ NO \quad Cl \end{bmatrix}_2 .$$

can be deduced.

These addition products differ in their behaviour from those of the tri-alkyl type $R_2C = CHR$. The latter show the same behaviour towards all reagents, whilst the nitrites of the type $RCH = CHR$ for example with sodium methylate yield the unsaturated nitro-compound $RCH = C(NO_2)R$, but the nitroso chlorides on the other hand yield substituted oximes:

$$R C \!-\!-\! CHR$$
$$\| \quad\quad |$$
$$NOH \; OCH_3$$

As compounds of the last-named type, anethol, isosafrol, and methyl isoeugenol were mentioned, for comparison with the previously examined phellandrene.

If anethol nitrite (melting point $121°$) is introduced into acetyl chloride, or treated with sodium methylate, β-nitro anethol is obtained in the form of a yellow oil which, after distillation *in vacuo*, is obtained as needles melting at $47°$. Its constitution is $C_6H_4(OCH_3)CH : C(NO_2)CH_3$. If hydroxylamine is allowed to act on this compound, anise aldoxime is split off; by heating with potash in alcoholic solution anisic aldehyde is formed. The diketone which, according to Toennies' formula

$$C_6H_4(OCH_3)C - COCH_3$$
$$\|$$
$$NOH$$

[1]) Liebig's Annalen **332** (1904), 305.

might be expected from nitro anethol, could not be obtained by splitting up with hydrochlorid acid. By means of reduction of the nitro compound with zinc dust and glacial acetic acid, the oxime $C_{10}H_{13}NO_2$ (melting point 65° to 66°) could be obtained, which, when split up with dilute sulphuric acid, is converted into the corresponding ketone

$$C_6H_4(OCH_3)CH_2 — CO — CH_3 \text{ (d 1,07; } n_{D20°} \text{ 1,5253),}$$
melting point of the semicarbazone 175°.

The anethol nitroso chloride (melting point 127° to 128°) obtained in the usual manner, when heated with alcohol or glacial acetic acid, yields anisic aldehyde; with sodium methylate, the oxime $C_{11}H_{15}NO_3$ (melting point 48°). If the latter, or better its sodium salt, is decomposed with 10 per cent. sulphuric acid at 80°, the ketone $C_{11}H_{14}O_3$ is obtained, whose semicarbazone melts at 192°.

$$\left[\begin{array}{c} C_6H_4(OCH_3)CH — CHCH_3 \\ \quad\; | \qquad\quad | \\ \quad\;\; NO \quad\;\; Cl \end{array}\right] \rightarrow \begin{array}{c} C_6H_4(OCH_3)C — CH — CH_3 \\ \qquad\qquad\quad \| \qquad\quad \diagdown \\ \diagup \qquad NOH \quad Cl \end{array}$$

$$\diagup \quad C_6H_4(OCH_3)CHO + NH_2OH \cdot HCl + CH_3CHO$$

$$\downarrow C_6H_4(OCH_3)C — CH \cdot CH_3$$
$$\qquad\qquad \| \qquad\quad \diagdown \qquad \rightarrow C_6H_4(OCH_3)COCH(OCH_3)CH_3$$
$$\qquad\;\; NOH \;\; OCH_3$$

Isosafrol nitrite (melting point 128°), when treated in a suitable manner, yields entirely analogous bodies as anethol nitrite: β-nitro-isosafrol, yellow crystals of the melting point 98°; with soda in alcoholic solution it splits off piperonal oxime, and can be reduced to the oxime $(CH_2O_2)C_6H_3 \cdot CH_2 \cdot C(NOH) \cdot CH_3$, which can be converted into the corresponding ketone (d 1,203; $n_{D20°}$ 1,5430). Melting point of the semicarbazone 163°. Isosafrol nitroso chloride, when treated with sodium methylate, yields the oxime

$$(CH_2O_2)C_6H_3 C — CH — CH_3$$
$$\qquad\qquad\quad \| \qquad\quad \diagdown$$
$$\qquad\qquad\;\; NOH \quad OCH_3$$

Analogous compounds are obtained under the same conditions from methyl isoeugenol.

———————

In a work published by E. Schmidt[1]) on the same subject, it is shown that the anethol nitroso chlorides obtained by Tilden and Forster's[2]) process are identic with those obtained by Wallach's[3]) method.

———————

[1]) Apotheker-Zeitg. **24** (1904), 655.
[2]) Berl. Berichte **27** (1895), ref. 467.
[3]) Liebigs Annalen **245** (1888), 251.

Walter Busse[1]) has published an interesting pharmacognostic study „On the medicinal and commercially useful plants of German East Africa". Without entering into a description of the many new medicinal and poisonous plants which the author has discovered during his long sojourn in the Colony, we will only refer here to some plants observed by him, which are worthy of attention on account of their content of essential oils and odorous substances.

Of several species of Ocimum growing wild along the roads and in the corn fields, *Ocimum canum* Sims. was especially met with; this species attracts attention on account of its refreshing aromatic odour, and is used in dwelling houses for driving off mosquitoes. Various other plants were noticeable for their far-spreading heliotrope odour; for example the Rubiacea *Plectronia heliotropiodora* K. Sch. n. sp., growing in the neighbourhood of Lake Nyassa, whilst a white-flowering Composita, probably *Veronica* sp. exhaled an even more penetrating heliotrope odour. The Rutacea *Clausena anisata* (Willd.) appears also to contain piperonal, contrary to the opinion of its discoverer who had described this plant as having an odour like anise. Finally a species of Ehretia (fam. *Borraginaceae)* of the island Kwale should be mentioned, whose blossoms also had a heliotrope odour, whilst Busse was unable to observe the heliotrope perfume in *Heliotropum strigosum* (Willd.) As a new coumarin plant with an exceptionally powerful aroma, the only species of Eupatorium hitherto become known from Africa, *E. africanum* Hiern, was found in the highlands of Ungoni; here, as in other cases, the odour only occurs when the plant is fading. As new representatives of eugenol-containing plants, there were found in the Sachsenwald and in the neighbourhood of Amani two labiates which have not yet been defined. Of the jasmine species *J. tettense* Kl. and *J. Bussei* Gilg were remarkable for their wonderful perfume. We would further mention *Asparagus racemosus* Willd. („mssanssa"), a white-flowering shrub with an elder-like odour, in Ostukami, and also various Leguminosae whose blossoms are conspicuous by their aroma, for example *Baphia cordifolia* Harms n. sp. from the Ugogo desert, *Baphia Kirkii* Bak., a tree of the littoral, much valued on account of its excellent wood, and *Baphia Busseana* Harms n. sp. found in Ungoni, which has a particularly powerful and pleasant odour.

E. Charabot and J. Rocherolles[2]) now give a connected report on their experiments respecting the distillation of terpene derivatives

[1]) Berichte d. deutsch. pharm. Ges. **14** (1904) 205.
[2]) Bull. Soc. chim. III. **31** (1904), 533.

with water vapours. Apart from a series of new examples of experiments, the work gives no fresh information, and we may therefore confine ourselves to a reference to what we said in our last Report[1]).

From the pen of Dr. Georg Cohn a monograph on „Die Riechstoffe" has been published by Fr. Vieweg & Sohn, Brunswick. It gives a review of the chemical individuals which have been isolated up to the present and which are important as bearers of perfumes, of their presence, isolation, production, properties, and tests, and deals with the result of the pharmacological and phyto-physiological research. Whereas this book is intended more particularly as an introduction in this domain, a treatise by Jeancard and Satie "Abrégé de la chimie des parfums" (Paris, Gauthier-Villars and Masson & Cie.) serves more as a short hand-book which naturally does not give any new information, but of which the value lies in the suitable arrangement of the matter, and in the tabular form of the summary of the most important constants.

Pharmacologico-physiological Notes.

A. J. J. Vandevelde[2]) has, by means of plasmolysis, compared a number of essential oils and odoriferous individuals which frequently occur as important components of the former, for their poisonous character. This was carried out by ascertaining a numerical value, the so-called critical coefficient, i. e. the number of grams of the substance under examination, which is isotoxic with 100 gm. absolute ethyl alcohol in its action on the living cell. The action of the alcohol in its turn was again compared with that of a solution of sodium chloride of a definite degree of concentration, which effects or removes the occurrence of the plasmolysis in the cell, and was reduced to 100 gm. alcohol. Consequently, the smaller this critical coefficient of a compound or of an oil is, the greater will be the toxic effect on the organism. Vandevelde now found that for alcohols this critical coefficient is largest; for aldehydes, phenols, etc. it is decidedly smaller. The bodies examined range themselves, according to the values obtained, in the following series, if the coefficient for absolute ethyl alcohol is taken as 100:

1. Thymol	0,04	5. Clove oil	0,21
2. Menthol	0,18	6. Thyme oil (white)	0,36
3. Cinnamic aldehyde	0,20	7. Ceylon cinnamon oil	0,44
4. Cassia oil	0,21	8. Thyme oil (red)	0,60

[1]) Report April 1904, 102.
[2]) Bull. de l'Assoc. Belge des chimistes 17 (1903), 269.

9.	Peppermint oil	1,52	19.	Angelica oil (root)		7,84
10.	Nutmeg oil	1,77	20.	Anise oil		10,33
11.	Star-anise oil	1,81	21.	Cognac oil (artificial)		13,83
12.	Carvone	2,06	22.	Anethol		17,08
13.	Bitter almond oil	2,26	23.	Lemon oil		18,55
14.	Benzaldehyde	2,26	24.	Cognac oil		18,89
15.	Caraway oil	5,40	25.	Isobutyl alcohol		21,20
16.	Lemon oil (terpene-less)	6,45	26.	Propyl alcohol (normal)		45,50
17.	Neroli oil	7,11	27.	Ethyl alcohol		100,00
18.	Carvene	7,71	28.	Methyl alcohol		142,10

This shows that the phenols and aldehydes (1 to 8) are at least 100 times more poisonous for the organism than absolute ethyl alcohol; the aldehydes, ketones, and terpenes (9 to 14) 20 to 100 times, the terpenes and ketones (15 to 20) 10 to 20 times, the esters (21 to 24) 5 to 10 times, and the alcohols up to 5 times more, or as in the case of methyl alcohol, less poisonous.

Bornyl valerianate. In recent times O. Engels[1]) has used bornyl valerianate in numerous cases of neurosis of the most varied character, including that of children, and he has been able to confirm its excellent effect on which L. Hirschlaff[2]) had already reported. It is especially useful in traumatic neurosis, hysteria, neurasthenia, epilepsy, nervous disorders of the stomach, and also in gynaecological cases. It may be mentioned particularly that it stimulates the appetite, and is free from all disagreeable secondary symptoms.

Menthyl valerianate. As already mentioned by us in our last Report[3]) menthyl valerianate has been found to be an excellent prophylactic against seasickness. K. Köpke[4]), in a treatise on seasickness, recommends this preparation on the grounds that — though not absolutely infallible, — it yet rarely fails to act. In the early stages of the sickness it is best taken in 10 to 15 drops on a lump of sugar. If this dose should not have the desired effect, it may be repeated after half an hour, with observation of the strictest diet.

Menthol. Contrary to his earlier statement, M. Bial[5]) has after later experiments established, that in the bile of dogs and cats, when

[1]) Therap. Monatsh. **18** (1904), 235.
[2]) Report April **1904**, 104.
[3]) Report April **1904**, 104.
[4]) Therap. Monatsh. **18** (1904), 296.
[5]) Centralbl. f. Physiol. **18**, 39. According to Chem. Centralblatt **1904**, II, 355.

these animals had received hypodermic injections of menthol dissolved in olive oil, a linked menthol compound appears, possibly menthol glucuronic acid.

_____

Since R. Kobert and E. Levy's experiments had shown that camphocarbonic acid and its esters are physiologically absolutely in-operative, the same scientists[1]) have also examined some acetyl deriv-atives of camphor for their action on the animal organism. It was then found that oxymethylene camphor behaves like camphocarbonic acid, and has no camphor-action whatever.

Its bactericide properties are also very slightly developed. Oxy-ethylidene- and oxypropylidene camphors, however, very much resemble laurus camphor, in their physiological action. R. Gottlieb observed on rabbits, after hypodermic injection of the ethylidene compound, a specially rapid and powerful spastic effect. The propylidene com-pound is also a spasm-producing toxin, though less powerful in its action.

Camphor mono- and di-iodide have been tested for their phy-siological behaviour by Messrs. E. Merck. The former has no decided antiseptic properties, and only irritates sensitive mucosa. It does not give rise to camphor-symptoms. It is non-toxic. Camphor di-iodide is a more powerful irritant than camphor iodide. It has a caustic action on the mucous membrane of the skin and of the stomach. It may possibly be found useful in cases where, in addition to an irritating effect, an absorbing action is desired, for instance in all chronic eczemata and similar affections.

Physiology of the sense of smell.

H. Zwaardemaker publishes some interesting treatises on the physiology of the sense of smell. Before entering into details, the olfacto-physiological measuring apparatus (olfactometer) and the unit of measure (olfactia) are described. The olfactometer[2]) consists essentially of a smelling cylinder (magazine cylinder), and a tube through which the smelling takes place. The magazine cylinder is pushed either completely or only partly over the tube, according to the requirements of the test, and by these means the limits of excitement and of perception are ascertained, that is to say, the position in which an undefined or a qualitative olfactory sensation occurs. In an olfacto-meter, whose magazine cylinder consists of india-rubber, there occurs for example, for a normal organ of smell, an olfactory sensation, when 0,7 cm. of the india-rubber cylinder is not covered by the inner tube.

_____

[1]) loc. cit.
[2]) H. Zwaardemaker, The physiology of the sense of smell, Leipzig 1895.

This value is then selected as the physiological unit of measure; it is the normal "minimum perceptibile", or the "olfactia".

A counterpart to olfactometry is odorimetry. For the latter, it is not a question of measuring the acuteness of the sense-organ, but of determining the intensity of the odour as an exciting agent of the sense. In this case also the unit of measure is the "olfactia".

An improvement of the olfactometer[1]) which is of importance for the olfacto-physiological measurements, consists of this, that a glass cylinder is attached to the smelling tube, and the former connected with a water jet pump of which the suction power can be controlled at any moment. By shifting the position of the magazine cylinder and drawing through air during 15 seconds, the glass cylinder (smelling flask) is filled with scented air, and the limits of excitement and perception determined by smelling tests. The physical measure is represented by an evaporating surface of known area, which is exposed to a current of air of constant and known velocity.

According to this fairly accurate physical measuring process, Zwaardemaker has submitted "the odorimetry of solutions of different percentages and of systems of heterogenous equilibrium"[2]) to a fresh examination.

The magazine cylinder consisted of filtering paper wrapped in several layers over a cylinder of copper gauze, and saturated with the odorous liquid to be examined. Experimental tests respecting the odorimetry of solutions of known percentages were first made by the old method, by smelling in the smelling tube, with aqueous solutions of camphor, coumarin, and ionones. A few of the results of these tests follow here:

1. β-ionone, aqueous solution with $1/2\,^0/_0$ antifebrin added.

$1:10^6$	limit of perception			0,5 to 1 cm.
$1:250000$	„	„	„	0,1 cm.
$1:10^5$	„	„	„	0,1 „
$1:10^4$	„	„	„	0,1 „

When the cylinder was pushed out further, the odour-intensity of the solution $1:10^4$ increases slightly, until 3 cm. is reached. From 6,5 cm. the odour acquires a herb-like character.

2. Ionone in aqueous solution $(1/2\,^0/_0$ antifebrin) $1:250000$.

α-ionone	limit of perception			0,05 cm.
β-ionone	„	„	„	0,1 „
irone	„	„	„	0,1 „
isoirone	„	„	„	2,1 „
ionone techn. a	„	„	„	2,5 „
ionone techn. b	„	„	„	0,1 „

[1]) Archiv für Laryngologie **15**, Vol. 2, 1.
[2]) Untersuchungen aus dem physiologischen Laboratorium 5. Series IV, II, 387.

· Of these cylinders, in the above arrangement of experiments, β-ionone and isoirone have a pleasant violet odour, but only in the vicinity of the limit of perception, at most up to an intensity of 5 olfactiæ; all other cylinders have a herb-like odour.

When diluted to $1:10^6$ all ionones have equal odour-intensity (0,1 cm.) and the violet-character becomes distinctly noticeable.

More exact measurements were taken with the above-described aspiration method. As it appeared desirable in the production of the perfume-solutions, to start from a physically exactly defined condition, which in theory has a more definite value than any arbitrary degree of dilution, a so-called heterogenous equilibrium was arranged on the magazine cylinder, by leaving the latter for a prolonged time in contact with water and camphor or some other odorous substance. There is then a system of two components (water with camphor), and three phases (solid, liquid, and gaseous); provided the temperature and the height of the barometer do not vary too much, the composition of the phases remains constant. As soon as the cylinder is shifted, and the continuous current of air disappears from its internal surface, the gaseous phase is removed. The evaporation of water and odorous substance, however, always takes place in the same proportion. If, therefore, the composition of the gaseous phase has previously been determined, it is possible to deduce from it the composition of the vapour which is liberated in the olfactometer. It is only necessary to calculate its degree of concentration from the total rapidity of evaporation of the liquid. The concentration of the odorous substance in the smelling flask is so great in this method of testing, that dilution is necessary for the purpose of determining the limit of perception. This is best done if necessary by repeatedly evacuating the smelling flask, and letting inodorous air pass through.

A few examples are given below:

1. Camphor water. System of two components and three phases. Approximate composition of the saturated solution of camphor $1:10^8$, at $12°$.

> Limit of perception at 0,7 cm./10 = 0,07 cm.
> „ „ „ „ 3,0 cm./16 = 0,19 „
> „ „ „ „ 6,0 cm./20 = 0,3 „

The camphor vapours consequently have a comparatively more powerful odour when in greater concentration, than when diluted.

2. β-ionone, $1^0/_{00}$ alcohol. Solution brought to an assumed system of equilibrium by means of water (three components, and two phases).

> Limit of perception at 2 cm./100 = 0,02 (faint violet odour).
> At 2 cm./20 = 0,1 cm. (faint herb-odour).

The advantage of producing a heterogenous equilibrium lies in this, that the passing of the odorous substance is proportionate to the size of the contact-surface, i. e. the pushed out length of the cylinder, and can be determined from the composition of the gaseous, and the rapidity of evaporation of the liquid phase. In this manner odorimetry, designed in the first instance as an indirect method, now develops into a direct one. It shows us at once the quantity of odorous substance which must be added to the unit of volume of inhaled air in order to produce a sensation of smell; but this would only be of importance for odorimetry in itself, if the measure of olfactive energy which is just sufficient for exciting the sense-organ, could be measured.

A further examination deals with the subject of „tasting by smell"[1]), an occurrence which is especially clearly noticeable in the form of a sweet taste, which can be brought about by the odour of chloroform. Zwaardemaker believes that the seat of this excitement of the sense may be localised in the epithelial buds recently discovered in the regio olfactaria of mammals. By means of olfactometric measurements it was found that the limit of excitement of the odour-sensation requires 2,6 mg. chloroform in 1 litre air, that of the taste-sensation 13,0 mg.

Of particular interest is a treatise on the „sensation of inodorousness"[2]).

Absolutely inodorous surroundings are in Zwaardemaker's opinion, like an absolutely dark and noiseless space, of rare occurrence, as almost every substance possesses its specific odour. The fact that we do not usually perceive it, depends, apart from our carelessness, upon the ventilation which enfeebles the odour, and, perhaps more correctly, produces transitions. If, therefore, we desire to become acquainted with the sensation of inodorousness, we have generally to resort to an intentionally produced state of inodorousness. The most suitable arrangement for this purpose is a glass case of 40 cc. capacity with an opening for the nose. If this case is carefully freed from adhering odours, the air contained in it will be found nearly odourless.

A different kind of inodorousness which is more often realised, consists of the compensation of odours which supplant each other. A complete compensation, however, occurs only in the case of very faint odours, whilst powerful excitants enter into a contest with one another. A difference between the sensation of this apparent inodorousness and the actual one does not exist. Zwaardemaker also considers a third kind of inodorousness possible. Concentrated solutions of many odorous substances, as compared with dilute ones, have a remarkably faint odour. On repeated observation it becomes more and more faint, and finally disappears almost completely. It is not

---

[1]) Arch. f. Anatomie und Physiologie **1903**, 120.
[2]) Untersuchungen aus dem physiologischen Laboratorium 5. Series IV, II.

improbable that this kind of inodorousness also occurs in nature. In the summer, for instance, the terpene odour of a fir-wood is not so striking in the wood itself as when such a wood is passed on an open road and the wind wafts the odour towards one.

So far we have only discussed the question of inodorousness in spaces, or of the air in the open. There is, however, also a condition of inodorousness of bodies. A body is usually understood to be inodorous, if no molecules at all, or only an infinitesimal number, become detached from the body by evaporation. A glass or platinum vessel is for this reason practically odourless. H. Erdmann[1]), however, had come to a different conclusion on the strength of his observations respecting the behaviour of odorous substances towards liquid air. Hanny and Horgarths have demonstrated that the capability of dissolving, which liquids possess towards non-volatile bodies, or towards such which only volatilise with difficulty, is also found in gases. Starting from this idea, Erdmann examined liquid air as a solvent, and found that three well-known odorous substances, citral, rose oil, and ionone, show a remarkably strong solubility in liquid air, and the author believes that this is a characteristic property of odorous substances. That which is usally called evaporation might therefore also be designated as solution in gaseous air.

Up to the present it has been accepted that the molecules which volatilise or dissolve in the air, belong to the bodies which chiefly build up the respective substances. In many cases occurring in nature this is not so, for example, in the case of a resin or a wax, in which the odorous particles only represent a small portion. From Erdmann's point of view the separation of odorous molecules is there simply a transition from one solvent into another. In such case the criterion is not the volatility, but the "coefficient of dissolution". This may be elucidated by the following example:

A one per cent. solution of β-ionone in *Paraffinum liquidum* is absolutely odourless, although a very large quantity of the odorous substance is present in the liquid. It is therefore necessary to imagine that the coefficient of dissolution between air and paraffin, for β-ionone, is absolutely in favour of paraffin. If the paraffin solution is shaken with water, the water acquires the most beautiful violet odour. Consequently, the coefficient of dissolution is more favourable towards water. Many bodies are therefore odourless, not because they do not possess the volatility of the odorous constituents, but because the coefficient of dissolution between the latter and the air is exceptionally unfavourable, or, in other words, because the solvent does not permit any appreciable vapour-pressure of the odorous constituent.

---

[1]) Journ. f. prakt. Chem. II. **61** (1900), 225.

Further, there are a number of substances which are very volatile, and which according to their chemical structure belong to the odorous substances, and yet do not smell. For such cases a plausible explanation may perhaps also offer itself. The air charged with the substance in question, when inhaled, comes in contact with the olfactory cells. The latter possess in their hair-like attachments a considerable enlargement of their free surface, and an extended contact takes place therefore between the air and the substance of the olfactory hairs. There will once more occur an interchange of the dissolved constituents between contiguous solvents. The quantitative conditions are again governed by a coefficient of dissolution. If the latter is unfavourable towards a transition into the substance of the olfactory hairs, then the molecules, even when they for themselves have a strong odour, will produce no excitement.

Finally we will mention two experiments[1]: intermission of the olfactory excitement by means of periodical interruption during $1/4$ of a second of the current of air, 1. inside the nose, and 2. outside, in the olfactometer. The two experiments have opposite results. In the former case, when the interruption takes place in the nasal cavity, the sensation is intermittent; in the latter, when the interruption is effected in the olfactometer, it is continuous. Zwaardemaker therefore concludes that the solution of the odorous molecules in the substance of the olfactory hairs takes place instantaneously, and that the fusion of the olfactory excitement does not originate in the terminal apparatus of the sense organ as such, but on the contrary in the air passages leading to it.

Phyto-physiological Notes.

Experiments made by E. Charabot and A. Hébert[2]) with regard to the acid-content of plants, show that the leaf, i. e. the organ of assimilation in which the hydrocarbons are worked up, is richest in free volatile acids. The content of volatile acids differs for each individual organ; the same applies to the individual stages of development. For example, it diminishes particularly when the inflorescence is in course of development; it increases during the unfolding of the blossoms, and finally again decreases. In the case of the geranium and the sweet basil plants, and also in the green organs of the sweet and bitter orange trees, the conditions are wholly identical. In the orange blossoms, however, the content of volatile acid is larger than in the stalk. Plants grown in the dark carry a greater proportion of acid in the roots than in the leaf. The suppression of efflorescence effects an increase of the acid-content in the leaf, to the disadvantage of the other organs.

---

[1]) Archiv für Anatomie und Physiologie 1904, 43.
[2]) Compt. rend. 138 (1904), 1714.

Hydrocarbons.

Pinene. As long ago as 1877 Tilden pointed out that optically strongly active pinene gives inferior yields in the production of nitroso-chloride. As pinene nitrosochloride (as mentioned by v. Baeyer) is bimolecular, Tilden[1]) believes that the small yield of nitrosochloride in the case of highly rotatory pinene must be attributed to the inversion process of the one half of the hydrocarbon, but that, on the other hand, with an inactive mixture of d- and l-pinene the bimolecular compound is only formed by the combination of the two semimole-cules. Experiments have confirmed the correctness of this view. American oil of turpentine with a rotatory power of $+7°$ in a 200 mm. tube, yielded 31 to 32 per cent. nitrosochloride; l-pinene ($a_D - 57°48'$ in a 200 mm. tube) 20 %, and d-pinene ($a_D + 80°41'$ in a 200 mm. tube) only 5,5 %. On the other hand, an inactive mixture of the two last-named bodies yielded 55,6 %. The production of the nitrosochloride was carried out as follows, contrary to the method hitherto in vogue: pinene is diluted with two to three times its volume of petroleum ether (boiling point 90° to 100°), cooled to 0°, and a solution of nitrosylchloride saturated at 0° in equal parts of petroleum ether and chloroform (containing about 8 % NOCl) added, whilst thoroughly stirred. When the reaction is completed the mixture is diluted with one and half times its volume of alcohol, and the crystalline pre-cipitate filtered off after a short time. — In one experiment, when it was attempted to produce the nitrosylchloride not separately, but by the introduction of nitrous acid in concentrated hydrochloric acid, which had been overlaid with pinene and petroleum ether, the formation of i-carvoxime hydrochloride (melting point 127° to 128°) was observed. — The earlier statements as to the melting point of pinene nitrosochloride are, according to Tilden, incorrect. The melting point of the com-pound washed with alcohol and dried at 50°, lies at 109° to 111°; after recrystallisation from chloroform, when preferably only small quan-tities are worked up at a time, as otherwise decomposition into nitroso-pinene (melting point 129°) occurs, the melting point lies at 115°. If pure pinene nitrosochloride is dissolved in chloroform, and alcohol is added, the solution acquires after some time an orange-red colour, reduces Fehling's solution, and contains clearly hydroxylamine. — Attempts to split up nitrosopinene and pinylamine into optical isomers, gave negative results. For the regeneration of pinene from the nitroso-chloride, methyl and dimethyl aniline, and dimethyl-p-toluidine are the most suitable agents. From the regenerated pinene, 55 % nitrosochloride were obtained. The author has finally determined the molecular weights

---

[1]) Journ. Chem. Soc. 85 (1904), 759.

of pinene nitrolbenzylamine, nitrolpiperidide, and nitrosocyanide, from which it would appear that these compounds are monomolecular. The nitroso-group in'these compounds assumes an isonitroso or oxime structure.

Camphene. G. Wagner, St. Moycho, and F. Zienkowski[1]) report on some experiments made to clear up the constitution of camphene. In continuation of earlier experiments by Wagner, the authors have submitted comparatively large quantities of pure camphene to oxidation with potassium permanganate. At ordinary temperature the action of 1 per cent. and 4 per cent. solutions of permanganate (1 atom oxygen to 1 molecule camphene) proceeds but very slowly. Comparatively many neutral and only few acid products are formed, and the bulk of the camphene remains unattacked. More rapid and complete is the oxidation with 4 per cent. permanganate solution (3 atoms oxygen to 1 molecule camphene) at 60°. Qualitatively the same products are formed, but the proportion of the acid and neutral constituents is reversed. In addition to the already known oxidation products: camphenylone $C_9H_{14}O$, campheneglycol $C_{10}H_{18}O_2$, camphene-camphoric acid $C_{10}H_{16}O_4$, and camphenylic acid $C_{10}H_{16}O_8$, a neutral compound $C_{16}H_{10}O_2$ of the melting point 169° to 170° was isolated (from ether-ligroin). The nature of this compound has not yet been cleared up. On oxidation with permanganate there are formed one ketone, whose semicarbazone melts at 184,5°, and two acids $C_{10}H_{14}O_8$, melting point 197° to 198°, and $C_{10}H_{16}O_4$, melting point 203°. As the oxidation-products differ from those of camphene glycol, the two compounds appear to have totally different constructions. For pure camphene-glycol, the melting point 199° to 200° was ascertained. During the treatment of camphene with permanganate, a small portion escapes oxidation. After suitable purification, a hydrocarbon $C_{10}H_{16}$ was isolated from it, which had the following constants: melting point 67,5° to 67,8°; boiling point 152° to 153° (757,5 mm. pressure). It appears to be identic with Wagner and Godlewski's cyclene.

In order to verify the correctness of Wagner's[2]) formulæ for camphenylone (I) and isoborneol (II), Tschugaeff had attempted to obtain isoborneol by the action of magnesium and methyl iodide on camphenylone

but he had thereby found that the alcohol which is formed during this reaction, is not identic with isoborneol. According to tests by the

---

[1]) Berl. Berichte **37** (1904), 1033.
[2]) Journ. russ. phys.-chem. Ges. **29** (1897), 121.

authors, this alcohol boils at 204° to 206° with decomposition, and sublimes *in vacuo* already below its boiling point. By crystallisation from ligroin it is obtained in the pure state. Its melting point lies at 117,5° to 118°; that of its phenyl urethane at 127,5° to 128°. On acetylation with glacial acetic acid and sulphuric acid, according to Bertram and Walbaum's method, there was formed, in addition to camphene, an acetate boiling at 98° to 103° ·(15 mm. pressure) which on saponification yielded an alcohol melting at 191° to 195°, presumably a mixture of isoborneol and the new alcohol; the formation of the isoborneol explains itself plainly.

Limonene. Tilden and Leach publish some communications on limonene nitrosocyanide[1]). As had been shown already previously, it is formed by the action of potassium cyanide on limonene nitrosochloride. Limonene-β-nitrosocyanide is soluble in alcohol, ether, and benzene, and is obtained from petroleum ether in the form of colourless prisms melting at 90° to 91°. It is, like the nitrosochloride, optically active, $\alpha_D \pm 165°$ in benzene solution. Its benzoyl derivative, when recrystallised from alcohol, forms tablets of the melting point 107°. Limonene-β-nitrosocyanide is monomolecular. Having regard to the limonene formula recently confirmed by Perkin, the following constitution belongs to it:

$$CH_3 \cdot C \underset{\underset{CN}{|}}{\overset{\overset{NO}{|}}{\big<}} \begin{matrix} CH - CH_2 \\ CH_2 - CH_2 \end{matrix} \big> CH \cdot C \big< \begin{matrix} CH_2 \\ CH_3 \end{matrix}$$

When α-nitrosochloride is treated with potassium cyanide, a product is formed which contains, in addition to carvoxime (melting point 72°), a liquid nitrosocyanide, which, however, on benzoylising yields the same derivative as β-nitrosocyanide. For this reason the two cyanides appear to be identical.

Thujene. J. Kondakow[2]) has shown that the haloid anhydrides of thujyl alcohol are mixtures of tertiary and secondary compounds, of which the former readily split off haloid hydracids, and yield two isomeric hydrocarbons. One of these is thujene, a true dicyclic hydrocarbon, identic with one of the products formed from thujyl methyl xanthogenate. On the strength of these results, J. Kondakow and V. Skworzow[3]) now also consider thujylchloride as a mixture of

[1]) Journ. chem. Soc. **85** (1904), 9.
[2]) Chem. Ztg. **26** (1902), 720. — Report October **1902**, 92.
[3]) Journ. f. prakt. Chem. II. **69** (1904), 176.

a secondary and a tertiary compound. They have recently gone further into this question by examining the behaviour of the above hydrocarbons and of sabinene towards haloid hydracids, and they have now arrived at the conclusion that the occurrence of liquid dihalo-hydrogen compounds, which are particularly formed during this treatment from thujene (obtained from thujyl haloid anhydrides), must be attributed to the high-boiling portions mixed with thujene, i. e. iso-thujene (Semmler's tanacetene). For this reason the solid dihaloid anhydrides which can be converted into trans-terpin derivatives, are formed from the true dicyclic thujenes. Isothujene can be recovered from the liquid dihalogen compounds, by splitting off two molecules haloid hydracid; but it does not appear to be identic with the tana-cetene, thujene, and isothujene from the amines. Possibly it is here a question of mixtures of isomerides.

Now with regard to the formation of the dihalogen compounds of thujene (from the xanthogenate), it may possibly be explained most simply by accepting the presence of admixed isothujene. The latter may be formed by isomerisation from thujene. To the last-named, the dextrorotatory hydrocarbon with lower boiling point and smaller specific gravity, the authors give the following constitution:

$$CH_3 \cdot CH \diagdown \diagup \begin{matrix} CH = CH \\ \diagdown \end{matrix}$$

whilst the following formula would then belong to the isomeride:

$$CH_3 \cdot C \diagdown \diagup \begin{matrix} CH - CH_2 \\ \diagdown \end{matrix}$$

On oxidation with potassium permanganate, the authors obtained from the first-named more stable hydrocarbon, homotanacetone dicar-boxylic acid of the melting point $146°$ to $147°$; from the isomeride with the higher boiling point, i. e. the isothujene, they obtained a- and β-tanacetogene dicarboxylic acids (melting point $141°$, and $116°$ to $117°$). Later in his work, Kondakow refutes the arguments which Tschugaeff[1]) and Semmler[2]) bring forward against Konda-kow's thujone formula, and endeavours to explain by means of his formula the conversion of tanacetone into carvotanacetone, and also that of a-tanacetone ketocarboxylic acid into the β-form.

Contrary to Kondakow and Skworzow, Tschugaeff[3]) explains in a subsequently published work on thujone derivatives, the formation

[1]) Chem. Ztg. **27** (1903), 970.
[2]) Berl. Berichte **36** (1903), 4367.
[3]) Berl. Berichte **37** (1904), 1481.

of the isomeric thujenes by accepting two stereo-isomeric thujyl alcohols, or xanthogene derivatives of these alcohols. Having regard to the acids obtained by Kondakow and Skworzow on oxidation of the isomeric thujenes, Tschugaeff points out that these results can be explained in the most natural manner by means of the constitutional formula drawn up by him for α- and β-thujone. He has further produced from thujamenthol an optically inactive hydrocarbon $C_{10}H_{18}$, of the following physical constants: $d\frac{20^\circ}{4^\circ}$ 0,8046, n_{D20° 1,44591, boiling point 157° to 159° (750 mm. pressure). From these constants Tschugaeff concludes that this body may possibly contain a pentamethylene ring, and that therefore the following constitutional formula may belong to it:

A crystalline nitrosochloride was obtained. The examination is to be continued.

With reference to the passage contained in Tschugaeff's above work, that he was the first to call attention to the formation of isomeric thujenes from thujyl xanthogenates, and that Kondakow had confirmed his results, the latter[1]) publishes a correction. He emphasises that he had first of all, and repeatedly, stated, that thujene and other hydrocarbons which are produced by means of the xanthogenate method, are not uniform, whilst Tschugaeff had always laid stress on their uniform character. If Tschugaeff now also stated that the thujene from the uniform thujyl xanthogenate was not uniform, this would be owing to the fact that he was compelled to do so by the force of the arguments brought forward by his opponents. For this reason the claims to priority made by Tschugaeff were not justified.

———————

Menthene. A few years ago we reported on Tschugaeff's[2]) method of producing hydrocarbons from xanthogene derivatives of terpene alcohols. Tschugaeff[3]) now has been able to produce and compare a number of menthenes from a whole series of such deriv-

———————

[1]) Journ. f. prakt. Chem. II. **69** (1904), 560.
[2]) Report April **1900**, 53 and October **1902**, 95.
[3]) Journ. russ. phys.-chem. Ges. **35** (1904), 1116. According to Chem. Centralbl. **1904**, I, 1347.

atives of menthol, and he has found that they are all identic. They have all the same boiling point (between 167° and 168°) and the same rotatory power $[a]_D + 113,28°$ to $116,74°$; the specific gravity is $d\frac{20°}{4°} 0,8122$, and the index of refraction $n_{D20°} 1,45242$. All the menthenes produced yield the same nitrosochloride. On fractional oxidation of all the hydrocarbons, only fractions of the same rotatory power are obtained, which proves their uniform character. With hydrogen bromide in solution of glacial acetic acid, the menthene from the xanthogene compounds yields a tertiary bromide which is optically almost inactive, p-4-bromomenthane, $C_{10}H_{19}Br$. To this belongs the structure of a p-menthene (3). From the nitrosochloride nitrosomenthene was obtained, from the latter menthenone, and from this by reduction menthol, so·that the cycle l-menthol, xanthogenic acid ester, menthene, nitrosochloride, nitrosomenthene, menthenone, l-menthol is complete.

Tschugaeff next discusses the menthenes which are obtained from l-menthol by means of halogen derivatives or by abstraction of water, and arrives at the conclusion that the menthene from halogen compounds consists chiefly of menthene (3), which yields a solid nitroso-chloride, and that the one obtained by splitting off water is a men-thene (4?) which does not yield a solid nitrosochloride.

Zelikow[1]) has used organic acids such as oxalic acid, succinic acid, tartaric acid, citric acid, phthalic acid, terephthalic acid and camphoric acid, as water-abstracting agents for the production of menthene from menthol. With the exception of tartaric acid, all the acids examined yielded the looked-for menthene. The best results were obtained with cam-phoric acid. The reaction takes place with formation of an acid ester, and decomposition of the latter when the temperature is raised further. In this process the presence of free acids plays an important part, as in such case a larger quantity of menthene is produced than in their absence. The dehydration was carried out by heating the mixture of menthol and acid in a retort in an air-bath. For details of the experiments with the various acids we must refer to the original work. The intermediate products isolated for the purpose of clearing up the reaction were, in addition to menthol, neutral and acid esters. The former, however, can play no part in the production of menthene, as for example neutral oxalic ester distils at 225° without decomposition, whilst the dehydration already takes place at 120°. If a large excess of free oxalic acid is present, menthene is no doubt formed during the distillation, but only in small quantity. The acid oxalic acid ester also, when distilled as such, yields no menthene, but it does yield the theoretical quantity when free oxalic acid is added.

---

[1]) Berl. Berichte **37** (1904), 1374.

Similar experiments were made with succinic acid, citric acid, phthalic acid, and camphoric acid, when the same results were obtained. The co-operation of the free acids in the dehydration process also explains why tartaric acid does not dehydrate menthol. As tartaric acid itself when heated changes considerably, it can no longer act on the ester formed.

Dry distillation of the sodium, barium, or calcium salts of the acid esters yielded also menthene, in addition to a small quantity of menthol.

M. Konowaloff[1]) describes some nitrogen compounds of the menthane-series. Menthane, produced by reduction of menthyl-bromide with zinc and hydrochloric acid in alcoholic solution (boiling point $169,5°$ to $170°$, $d_{0°}^{20°}$ $0,7929$, $n_{D21°}$ $1,43757$) can. be nitrated with dilute nitric acid both in open vessels and in closed tubes. The nitration-product examined up to the present, mononitromenthane

$$(CH_3)_2 C(NO_2) CH \left\langle \hexagon \right\rangle CH — CH_3$$

(boiling point $135°$ to $137°$, $d_{0°}^{22°}$ $0,9871$, $n_{D22°}$ $1,46241$) can be reduced with. tin and hydrochloric acid to the corresponding menthyl-amine (boiling point $199°$ to $200°$, $d_{0°}^{2°}$ $0,8451$).

Sesquiterpenes. Oswald Schreiner[2]) has in two further publications completed his monograph of the sesquiterpenes. It would take us too long even were we only to enumerate here the numerous representatives of this class of bodies. We will only mention that Schreiner reports on no less than sixty different sesquiterpenes, of which, as the author already points out in his introduction, no doubt many are identic, or mixtures of several isomerides.

Terpenes (as Ladenburg[3]) and Kriewitz[4]) have shown) combine with paraformaldehyde into additional compounds of an alcoholic character. According to examinations made by P. Genvresse[5]), the reaction with sesquiterpenes proceeds in the same manner, though with greater difficulty and with smaller yield. By heating the components to $180°$ under pressure, they combine into compounds of the composition $C_{16}H_{26}O$. The corresponding derivative of caryophyllene represents a liquid boiling at $177°$ to $178°$ (15 mm. pressure).

[1]) Journ. russ. phys.-chem. Ges. 36 (1904), 237. According to Chem. Centralbl. 1904, I, 1516.
[2]) Pharmaceut. Review 22 (1904), 101, 146.
[3]) Berl. Berichte 31 (1898), 289.
[4]) Berl. Berichte 32 (1899), 57, Report April 1899, 49.
[5]) Compt. rend. 138 (1904), 1228.

d_{0° 0,997, n_D 1,508, a_D — $7^\circ 40'$ in 4,93 per cent. chloroform solution; the acetate boils at 185° (15 mm. pressure); d_{0° 0,9969, n_D 1,490; $a_D + 20^\circ 33'$ in 11,7 per cent. chloroform solution. The compound of clovene with formaldehyde boils at 170° (12 mm. pressure); d_{0° 1,001, n_D 1,6105; a_D — $7^\circ 12'$ in 6,03 per cent. chloroform solution. The cadinene derivative shows: boiling point 180° (15 mm. pressure); d_{0° 0,993; n_D 1,521; a_D — $17^\circ 54'$ in 7,6 per cent. chloroform solution. From the indices of refraction the molecular refractions can be calculated, which in the case of the first two derivates point to one, and in the last-named to two double-linkings.

Alcohols.

In order to ascertain the influence on the rotatory power exerted by the entrance of unsaturated radicals into the hydroxyl-group of optically active alcohols, A. Haller and F. March[1] have produced the allyl ethers of l-borneol, l-menthol, d-methyl cyclohexanol, and l-linalool, and determined the optical rotation of these hitherto unknown compounds. The ethers are obtained by the action of allyl iodide on the sodium alcoholates in a solution of toluol. Except in the case of linalyl ether, it was found that the rotatory power of the ethers, as compared with that of the alcohols, had increased considerably. The molecular refraction of the ethers corresponded well with the calculated figures. Here also the ether of linalool made an exception, inasmuch as it gave a value which was somewhat too low. Attempts made to produce by the same method the corresponding propyl ethers, failed completely. Similar differences in the behaviour of various halogen alkyls towards sodium compounds have also been already observed in other cases.

By the action of the propyl ester of p-toluol sulphoacid $CH_3 C_6 H_4 \cdot SO_3 \cdot C_3 H_7$ on menthol sodium, there was next obtained n a not completely pure state the propyl ether of menthol, which had a somewhat lower optical activity than the one of the menthol used.

Menthol. L. Wedekind and K. Greiner[2] have submitted to .a critical examination with regard to the grounds for their existence, the four compounds of menthol and formaldehyde which have been described up to the present:

    1. methylenedimenthyl ether[3] $C_{10} H_{19} \cdot O \cdot CH_2 \cdot O \cdot C_{10} H_{19}$,
    2. chlormethylmenthyl ether[4] $C_{10} H_{19} \cdot O \cdot CH_2 Cl$,

---

[1] Compt. rend. **138** (1904), 1665.
[2] Zeitschr. f. angew. Chemie **17** (1904), 705.
[3] Compt. rend. **128** (1899), 612. Bull. Soc. Chim. III. **21** (1899), 370.
[4] Berl. Berichte **34** (1901), 816. Pharm. Zeitg. **46** (1901), 32. German Patent No. 119008.

3. mentholformaldehyde compounds which are believed to be formed by the solution of formaldehyde gas or of trioxymethylene in liquefied menthol [1]),

4. a mentholformaldehyde compound $C_{10}H_{19} \cdot O \cdot CH_2OH$ which is said to be produced from menthol and aqueous solution of formaldehyde by boiling with dilute hydrochloric acid [2]).

It was then found that only the compounds 1 and 2 are well characterised chemical bodies, whilst compound 3 represents a mixture of formaldehyde and menthol, which is already evident from the fact that in an aqueous emulsion of the compound, the whole of the formalin can be determined as free formaldehyde. For compound 4 the following analytical composition was found:

free formaldehyde		0,19%
menthyl chloride	1,88 to	2,51 „
chlormethylmenthyl ether	5,1 „	6,5 „
free menthol	37,25 „	41,85 „
methylenedimenthyl ether	45,12 „	60,6 „

Citronellol. In our last Report we mentioned a method by Bouveault and Blanc [3]) for the production of primary alcohols from the corresponding acids, by reduction of the esters with sodium and alcohol. With the help of this process, Bouveault and Gourmand have successfully accomplished a complete synthesis of „rhodinol" $C_{10}H_{20}O$ [4]). The crude material employed was the ethyl ester of the synthetically approachable geranium acid, or better of dihydrogeranium acid. On reduction the carboxethyl-group is reduced into an alcohol-group, and moreover the β, γ-double-linking eliminated, whilst the ζ, η- (or the η, ϑ-) double-linking remains intact. The synthetic product proved in every respect identic with the alcohol $C_{10}H_{20}O$ present in rose oil and pelargonium oil. The identity of the two was confirmed by the production of the pyruvic acid ester, boiling point 143° (10 mm. pressure), and comparison of the semicarbazones of this ester, melting point 112°.

Bouveault designates the alcohol $C_{10}H_{20}O$ present in oils of rose and pelargonium, as „rhodinol", in contradistinction to citronellol, the reduction product of citronellal. To the former he ascribes the constitution:

---

[1]) Patent application D. 8876, cl. 12, o.
[2]) German Patent, as applied for St. 7278 IV, cl. 12, o.
[3]) Compt. rend. **136** (1903), 1676; **137** (1903) 60; Report April **1904**, 110.
[4]) Compt. rend. **138** (1904), 1699. The first complete synthesis of this alcohol, as is well known, was made by Tiemann. Berl. Berichte **31** (1898), 2899; Report April **1899**, 63.

$$\begin{array}{l} CH_3 \\ \!\!\!\searrow\!\!C = CH \cdot CH_2 \cdot CH_2 - CH - CH_2 \cdot CH_2OH \ \ \text{(rhodinol)}, \\ CH_3\!\!\nearrow \!\! | \\ CH_3 \end{array}$$

and to the latter

$$\begin{array}{l} CH_3 \\ \!\!\!\searrow\!\!C - CH_2 - CH_2 - CH_2 \cdot CH - CH_2 \cdot CH_2OH \ \ \text{(citronellol)} \\ CH_2\!\!\diagup\!\!\diagup \!\! | \\ CH_3 \end{array}$$

Bouveault[1]) supports this view by the difference in the behaviour of the aldehydes; Harries[2]) agrees with him, on the strength of the results obtained by him in the oxidation of the acetal of citronellal. As we happened to have at our disposal citronellol obtained by reduction from citronellal, we have produced from this alcohol the pyruvic acid ester in accordance with Bouveault's directions, and compared the semicarbazone of the latter with the corresponding compound from citronellol occurring in nature. For both semicarbazones we found the melting point at 110° to 111°, and a mixture of the two melted at the same temperature. In view of these results we cannot agree with Bouveault's opinion that the alcohol $C_{10}H_{20}O$ in rose oil and pelargonium oil is not identic with the reduction product of citronellal.

In recrystallising the semicarbazones from ethyl alcohol we found the melting point of the semicarbazones at 103° to 106°; if the melting point tube was dipped in sulphuric acid heated to 90°, the substance melted already slightly below 100°. Only after crystallising from methyl alcohol and drying in the vacuum desiccator we found the melting point mentioned above. It would appear that ethyl alcohol participates molecularly in the crystallisation, or at least is held very firmly mechanically, which would bring about the depression of the melting point.

Aldehydes and Ketones.

On the estimation of aldehydes and ketones.

An aldehyde-estimation by means of neutral sodium sulphite, lately proposed by Sadtler has already been discussed by us in our last Report (p. 48). This method consists of this, that the NaOH liberated by the action of sodium sulphite on aldehydes, is determined by titration, and from the result the aldehyde-content calculated. As it is said to give very accurate results, the method is above all re-commended for the citral-estimation in lemon oil. Sadtler also

---

[1]) Bull. Soc. chim. III. **23** (1900), 458; Report October **1900**, 77.
[2]) Berliner Berichte **34** (1901), 2081. Report April **1902**, 90.

mentioned that the method could be applied both to aliphatic and to aromatic aldehydes, and in a recently published work[1]) he discusses the subject still further. He found from subsequent experiments that the method is suitable for saturated and unsaturated aldehydes of the aliphatic as well as the aromatic series. Moreover, some ketones also react well with sodium sulphite, above all carvone, which quantitatively gives accurate results, whilst others, for example Japan camphor[2]), do not combine at all with sodium sulphite. Sadtler believes that in such cases double-linkings, or certain atom-groups, play a part.

The chemical processes which take place during the conversion are explained by Sadtler by the following formulæ:

1. for aldehydes:

$$R \cdot CHO + 2 Na_2 SO_3 + 2 H_2 O = R \cdot CH : (NaSO_3)_2 + 2 NaOH + H_2 O,$$

2. for ketones:

$$R \cdot CO \cdot R + 2 Na_2 SO_3 + 2 H_2 O = R_2 : C : (NaSO_3)_2 + 2 NaOH + H_2 O.$$

In our previous Report we already referred to the difficulties which titration offers in the case before us, and at the present moment we are not yet convinced of the suitability of this method for the citral-determination in lemon oil.

A further treatise, dealing with the same method of estimation, originates from H. E. Burgess[3]). His method differs from Sadtler's chiefly in the fact that the determination is purely volumetric. Burgess employs a small flask of 200 cc. capacity of which the neck carries a scale of 5 cc. divided in $\frac{1}{10}$, and which is moreover provided at one side with a tube reaching to the bottom of the vessel, for the introduction of oil, reagents, and water. In this flask 5 cc. oil are mixed with a saturated solution of neutral sodium · sulphite and two drops phenol phthalein solution, and heated on a water-bath. The mixture is repeatedly thoroughly shaken, and the liberated alkali neutralised from time to time with dilute acetic acid (1 : 10), until the red coloration has permanently disappeared. Water is then added to bring the oil in the neck of the flask, and when cold, the scale is carefully read. The quantity of oil absorbed, when multiplied with 20, gives the percentage of aldehyde or ketone.

It is claimed for Burgess' process that it has this advantage over the bisulphite method, that it takes up less time, and that the end of the reaction can be accurately observed. Moreover, it is said that

[1]) Journ. Soc. chem. Ind. 23 (1904), 303.
[2]) Sadtler designates it incorrectly as an isomer of carvone; the latter has the formula $C_{10}H_{14}O$, whilst Japan camphor has $C_{10}H_{16}O$.
[3]) The Analyst 29 (1904), 78.

no crystalline precipitates are hereby formed, whilst this is often the case with the bisulphite method, when it is impossible to read the meniscus accurately.

Burgess has made experiments with a whole series of aldehydes and ketones, and has thereby found that the following compounds give. useful results: benzaldehyde, anisic aldehyde, cinnamic aldehyde, citral, citronellal, cuminic aldehyde, nonylic aldehyde, decylic aldehyde, carvone, and pulegone. In the case of citronellal, nonylic, decylic and cuminic aldehydes, prolonged heating is necessary for a complete conversion; with cuminic aldehyde and pulegone, it is convenient to substitute litmus for phenol phthalein.

With the help of the method under discussion, Burgess has also ascertained the aldehyde- or ketone-content of various oils. We give here the results:

Cassia oil: 80 and 85% cinnamic aldehyde,
Ceylon cinnamon oil: 68, 72 and 74% cinnamic aldehyde,
Cumin oil: 24% cuminic aldehyde,
Dill oil: 50% carvone,
Spearmint oil: 62% carvone,
Caraway oil: 55 and 57% carvone,
Oil of European pennyroyal: 16% pulegone.

Whereas in most oils the determination of· aldehyde or ketone can be accomplished direct, Burgess recommends previous concentration in the case of citron oil, lemon oil, limette oil, and oil· of sweet orange, owing to their low aldehyde-content. For this purpose he takes 100 cc. oil, and at a pressure of at most 15 mm. distils off 90 cc.[1]), when the. residue is driven over with water vapour at ordinary pressure. The quantity of the steam-distillate is accurately determined, and. from this in the above manner the aldehyde-content estimated. By these means the citral - content ascertained was: in lemon oil about 3%, in pressed limette oil about 8%,·in citron oil about 4%, in oil of sweet orange about 0,75 to 1%.

We have up to the present not yet made a sufficient number of determinations by Burgess' method to form a conclusive opinion on the process, but according to our present experience it appears to be useful. Its principal value, as compared with the bisulphite method, is in our opinion, that it is capable of more general application. But we very much question the reliability of the method of citral-determination (the "indirect" determination) recommended for lemon oil etc., as the manner of concentration undoubtedly causes greater

---

[1]) In order to test the oil at the same time for adulteration with turpentine oil, the first 10 cc. of the destillate are collected separately, and examined for their optical behaviour.

loss than Burgess appears to accept, — a matter which is of considerable importance, in view of the low citral-content of the oils in question. We will return to this method when the opportunity arises.

Camphor. In several treatises J. W. Brühl[1]) provides contributions to the knowledge of the halogen derivatives of camphor. If oxymethylene camphor in alkaline solution is iodised, no camphor mono-iodide is formed, but a di-iodo-substitution product. In neutral solution iodo-formyl camphor, or iodo-camphoric aldehyde is formed, which when submitted to the action of alkali and iodine, is converted into the above o, o-di-iodo camphor. The latter, when treated with methyl alcoholic potash, with the application of heat, yields o-mono-iodo camphor. The last-named body is also formed when dilute sodium methylate, with application of heat, is allowed to act on iodo-formyl camphor. O-iodo camphor

$$C_8H_{14}\diagup\overset{CHI}{\underset{CO}{\vert}}$$

forms colourless prisms melting at $42°$ to $43°$. O, o-di-iodo camphor

$$C_8H_{14}\diagup\overset{CI_2}{\underset{CO}{\vert}}$$

entirely resembles in its appearance iodoform. It melts at $108°$ to $109°$.
Iodo-formyl camphor

$$C_8H_{14}\diagup\overset{CI-CHO}{\underset{CO}{\vert}}$$

crystallises in leaflets of the melting point $67°$ to $68°$. It is fairly readily decomposed. O-iodo camphor is also formed from o-bromo camphor, if iodine is allowed to act on the latter's magnesium compound:

$$C_8H_{14}\diagup\overset{CHBr}{\underset{CO}{\vert}} + Mg = \left| C_8H_{14}\diagup\overset{CH\cdot MgBr}{\underset{CO}{\vert}} + I_2 = C_8H_{14}\diagup\overset{CH\cdot I}{\underset{CO}{\vert}} + MgBrI\right.$$

further, from camphor, sodium amide, and iodine:

$$C_{10}H_{16}O + NaNH_2 = NH_3 + \left| C_{10}H_{15}NaO + I_2 = NaI + C_8H_{14}\diagup\overset{CHI}{\underset{CO;}{\vert}}\right.$$

and finally from camphor, sodium, and iodine:

$$C_{10}H_{16}O + Na = H + \left| C_{10}H_{15}NaO + I_2 = NaI + C_8H_{14}\diagup\overset{CHI}{\underset{CO.}{\vert}}\right.$$

In iodo-formyl camphor, both the formyl-group and the iodine atom can be readily substituted by hydrogen. Up to the present,

[1]) Berl. Berichte **37** (1904), 2156, 2163, 2178.

however, all attempts made by Brühl to convert this compound into formyl camphor or camphoric aldehyde, have failed. In every case only the isomeric oxymethylene camphor was obtained.

For the purpose of isolating the already known bromo-formyl camphor, the best method is the one mentioned above for iodo-formyl camphor. But it has not been possible to obtain from it the o, o-di-bromo product (melting point 61°) by the method indicated for o, o-di-iodo camphor. Only o-bromo camphor is formed; and only the latter is also obtained if oxymethylene camphor is treated with alkali and bromine, either the first or last named in excess. The conditions are precisely the same in the case of o-mono-bromo menthone and bromoformyl menthone. The former body was hitherto unknown. It is an almost colourless oil, which cannot be distilled without decomposition in vacuo, but on the other hand distils very well with water vapour. ───────

With reference to a patent of Messrs v. Heyden (French patent No. 339 504), relating to the production of camphor from oil of turpentine (by boiling with salicylic acid, saponification of the ester obtained, and oxidation of the alcohol formed), Tardy[1]) points out that already several years ago he obtained in the same manner esters of salicylic acid, and further borneol, and only abstained from communicating the fact, owing to the absence of technical uses for this reaction. ───────

Pulegone. If hydroxylamine is allowed to act on α, β-unsaturated cyclic ketones, there are formed under certain conditions, hydroxylamine addition products. The pulegone hydroxylamine thus formed has been examined by Semmler[2]) for its behaviour towards dehydrating agents. For example, the following two products might chiefly be produced by splitting off water:

[1]) Journ. de Pharm. et Chim. VI. **20** (1904), 57.
[2]) Berl. Berichte **37** (1904), 305.

Formula I could then immediately be converted, by disruption of the tri-ring, into a ketoalcohol, and the latter, by loss of water, yield an unsaturated ketone:

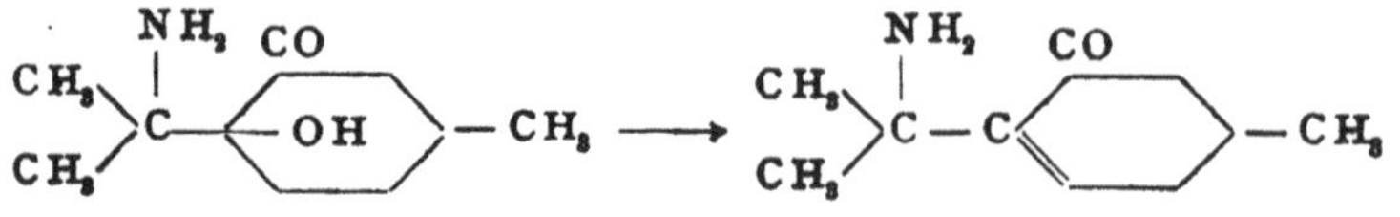

If pulegone hydroxylamine is dissolved in concentrated acids and heated to 50° to 100°, a mixture of bases is obtained, from which after purification a crystallising picrate can be obtained, which on decomposition yields the base $C_{10}H_{17}NO$, boiling point 91° (8 mm. pressure); $d_{20°}$ 0,9731; $a_D + 37° 10'$; n_D 1,4757. This body is a weak base, which gives nearly all alkaloid reactions. Its ketone character could be proved by formation of an oxime (melting point 180°) and a semicarbazone (melting point 153° to 154°). As its benzene sulpho-compound (melting point 120°) does not dissolve in alkalis, it is here a question of a secondary base. The hydrochloride could not be produced in the pure state, and for this reason no nitroso compound could be obtained. On reduction with sodium and alcohol, the base absorbs 4 atoms hydrogen and is converted into the tetra-hydro base $C_{10}H_{21}NO$, boiling point 134° to 135° (18 mm.); d 0,9646; n_D 1,4815. The latter does not yield a readily soluble picrate, but on the other hand, yields a carbamate. If the base $C_{10}H_{17}NO$ is reduced with zinc dust and hydriodic acid, only two atoms hydrogen are added, with formation of the base $C_{10}H_{19}NO$ (boiling point 106°, 11 mm. pressure); $d_{20°}$ 0,952; $a_D - 5°$; n_D 1,4727.

With regard to this work C. Harries and L. Roy[1] state that they have also obtained from pulegone hydroxylamine, by heating with 20 per cent. hydrochloric acid, the base $C_{10}H_{17}NO$.

For the further elucidation of the constitution of this body, Semmler[2] endeavoured to determine whether it is here a question of a dicyclic system, or whether after primary formation of the same, ring-disruption with formation of a double-linking has again taken place. Chiefly on account of the experimentally confirmed fact that the same above-mentioned tetrahydro base $C_{10}H_{21}NO$ is also obtained by reduction of two bodies whose constitution is established, viz. pulegone hydroxylamine and pulegone amine, he arrives at the conclusion that the above-named formula I belongs to the base $C_{10}H_{17}NO$. The following graphic formulæ illustrate what has been said:

[1] Berl. Berichte **37** (1904), 1341.
[2] Berl. Berichte **37** (1904), 2282.

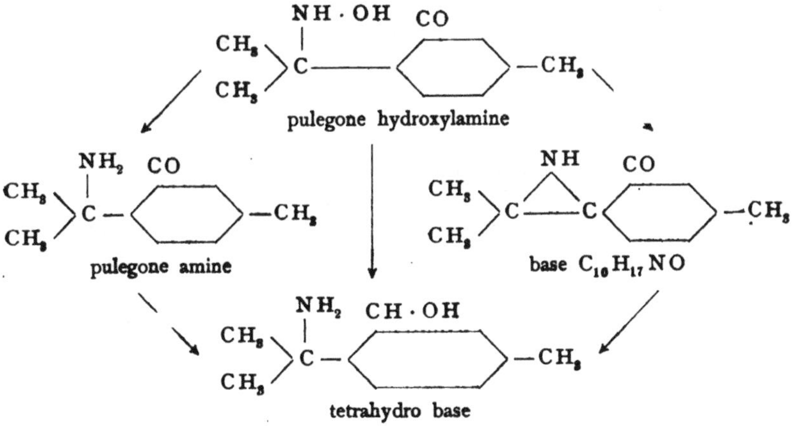

$$\text{CH}_3 \diagdown \atop \text{CH}_3 \diagup \text{C} \overset{\text{NH·OH}}{\underset{}{\Big|}} \overline{} \bigcirc \overline{} \text{CH}_3 \diagdown$$

pulegone hydroxylamine

$$\text{CH}_3 \diagdown \atop \text{CH}_3 \diagup \text{C} \overset{\text{NH}_2 \quad \text{CO}}{\underset{}{\Big|}} - \bigcirc - \text{CH}_3$$

pulegone amine

$$\text{CH}_3 \diagdown \atop \text{CH}_3 \diagup \text{C} \overset{\text{NH}}{\underset{}{\Big|}} \overline{} \text{C} \bigcirc - \text{CH}_3$$

base $C_{10}H_{17}NO$

$$\text{CH}_3 \diagdown \atop \text{CH}_3 \diagup \text{C} \overset{\text{NH}_2 \quad \text{CH·OH}}{\underset{}{\Big|}} - \bigcirc - \text{CH}_3$$

tetrahydro base

Menthone. For the production of alkyl- and alkylidene derivatives of cyclic ketones, A. Haller[1]) employed sodium amide instead of sodium metal hitherto used for this reaction, and by these means he prevented the formation of the corresponding alcohols which otherwise occur as by-products. By allowing aldehydes or halogen alkyls to act on the produced sodium compound of the ketone in question, the desired products of condensation or substitution were obtained directly. The alkyl menthones produced by Haller are colourless oils which, with the exception of methylmenthone, all have a musty and menthone-like odour. Up to the present the following bodies have been obtained:

methylmenthone boiling point 96° to 97° (13 mm. pressure) $[\alpha]_D$ +20,18°
ethylmenthone   „   „ 101° „ 102° (13 „   „   ) „ +38°
propylmenthone   „   „ 128° „ 132° (19 „   „   ) „ +39,20°
isobutylmenthone   „   „ 124° „ 128° (10 „   „   ) „ +45°
isoamylmenthone   „   „ 138° „ 143° (10 „   „   ) „ +31,48°
allylmenthone   „   „ 134° „ 137° (20 „   „   ) „ +25,42°

The anomalies of the foregoing compounds, with regard to boiling points and rotatory powers, which latter are all contrary to those of the used menthone ($[\alpha]_D$ —26,18°), may possible be attributed to the sensitiveness of the ketone towards alkalis and the formation of several stereo-isomerides. The synthesis will also be employed for camphor, pulegone, thujone, etc.

The menthone amine[2]) to be obtained by nitration of the menthone and subsequent reduction, yields on benzoylating two benzoyl-com-

[1]) Compt. rend. **138** (1904), 1139.
[2]) Chem. Centralbl. **1895**, II, 1160.

pounds which can be separated by crystallisation; with regard to the structure nothing definite can be stated. On reducing, aminomenthol is formed. Attemps to substitute in the latter the hydroxyl by bromine were not successful.

Carvone. Some time ago Paul Rabe und Karl Weilinger[1]) have demonstrated, that in attaching aceto-acetic ester to carvone by means of sodium ethylate, and subsequent elimination of the carbethoxyl-group, a dicyclic ketone alcohol, isopropylmethyl dicyclononanolone, is formed. In a second treatise[2]) on the subject in question, it is first of all shown that the same compound can be obtained from chloro-tetrahydrocarvonyl aceto-acetic ester (I), the condensation-product from aceto-acetic ester, carvone and hydrochloric acid. By the action of zinc dust, glacial acetic acid, and fuming hydrochloric acid on this substance, a mixture is formed of dihydrocarvonyl aceto-acetic ester (II) and the acetate of 8-oxyterpane-2-on-6-yl-aceto-acetic ester; melting point $133°$ from alcohol.

$$
\begin{array}{ll}
\text{I.} &
\begin{array}{l}
\text{COOC}_2\text{H}_5 \\
| \\
\text{HC}\!-\!\text{CH}\!-\!\!-\!\text{CH}_2 \\
|\quad\ |\qquad\quad | \quad\ \diagup\text{CH}_3 \\
\text{CO}\ \ \text{CH}\cdot\text{CH}_3\ \ \text{CHCCl}\!\!< \\
|\quad\ |\qquad\quad |\quad\ \diagdown\text{CH}_3 \\
\text{CH}_3\ \ \text{CO}\!-\!\!-\!\text{CH}_2
\end{array}
&
\begin{array}{l}
\text{COOC}_2\text{H}_5 \\
| \\
\text{CH}\!-\!\text{CH}\!-\!\!-\!\text{CH}_2 \\
|\quad\ |\qquad\quad | \quad\ \diagup\text{CH}_3 \\
\text{CO}\ \ \text{CH}\cdot\text{CH}_3\ \ \text{CH}\cdot\text{C}\!\!< \\
|\quad\ |\qquad\quad |\quad\ \diagdown\text{CH}_3 \\
\text{CH}_3\ \ \text{CO}\!-\!\!-\!\text{CH}_2
\end{array}
\\
& \text{II.}
\end{array}
$$

Dihydrocarvonyl aceto-acetic ester, a thick yellow oil, yields on intramolecular aldol condensation with simultaneous elimination of the carbethoxyl-group, a 1, 5-diketone, isopropenylmethyl dicyclonon-anolone (III),

$$
\begin{array}{ll}
\text{III.} &
\begin{array}{l}
\text{CH}_2\!-\!\text{CH}\!-\!\!-\!\text{CH}_2 \\
|\qquad\ |\qquad\quad | \quad\ \diagup\text{CH}_2 \\
\text{CO}\ \ \ \text{CH}\cdot\text{CH}_3\ \ \text{CH}\cdot\text{C}\!\!\Vert \\
|\qquad\ |\qquad\quad |\quad\ \diagdown\text{CH}_3 \\
\text{CH}_2\!-\!\text{COH}\!-\!\!-\!\text{CH}_2
\end{array}
&
\text{IV.}
\begin{array}{l}
\text{CH}_2\!-\!\text{CH}\!-\!\!-\!\text{CH}_2 \\
|\qquad\ |\qquad\quad | \\
\text{CH}_2\ \ \text{CH}\cdot\text{CH}_3\ \ \text{CH}\cdot\text{CH(CH}_3)_2 \\
|\qquad\ |\qquad\quad | \\
\text{CH}_2\!-\!\text{CH}\!-\!\!-\!\text{CH}_2
\end{array}
\end{array}
$$

a red-brown oil, which on reduction is converted into the solid glycol of the melting point $172°$ to $173°$[3]). If the glycol is heated with hydriodic acid to $220°$, there is formed 3-isopropyl-9-methyl dicyclo-[1, 3, 3] nonane (IV).

---

[1]) Berl. Berichte **36** (1903), 225. — Report April **1903**, 94.
[2]) Berl. Berichte **37** (1904), 1667.
[3]) Berl. Berichte **36** (1903), 230. — Report April **1903**, 94.

Methylhexanone. By treating methylhexanone (from pulegone) with sodium amide, and subsequently allowing alkyl iodides to act on the sodium compound of methylhexanone produced, A. Haller[1]) has obtained several homologues and one isomeride of menthone. The reaction takes place according to the following equation:

$$\text{H} \cdot \text{CH}_2 \underset{\text{H}_2 \quad \text{O}}{\overset{\text{H}_2 \quad \text{H}_2}{\hexagon}} \text{HNa} + \text{IR} = \text{H} \cdot \text{CH}_2 \underset{\text{H}_2 \quad \text{O}}{\overset{\text{H}_2 \quad \text{H}_2}{\hexagon}} \text{RH} + \text{NaI}$$

All the compounds obtained in this manner yield crystallised semicarbazones; they also combine with aromatic aldehydes, especially with benzaldehyde. In addition to dimethylhexanone, methylethylhexanone, methylpropylhexanone, and methylallylhexanone, Haller, conjointly with Martine, obtained in this manner an active menthone which is identic with the menthone obtained by oxidation from ordinary menthol. The boiling point lies at 207° to 209°; $a_D + 12° 56'$. The melting point of the oxime is 58° to 59°, that of the semicarbazone 183° to 184°.

O. Wallach[2]) communicates a new case of optical isomerism which has become known on the occasion of a work on methylhexanone. The methylhexanone obtained by splitting up pulegone is dextrorotatory. It has the specific rotatory power $[a]_{D\,15,8°} + 13,33°$. In solution, the rotation depends on the character of the solvent. For instance, with alcohol it is $[a]_{D\,14°} + 10,45°$; with ether $[a]_{D\,15°} + 17,44°$. Pure R-l-methylhexanone oxime, obtained from R-methylhexanone, whose uniform character was demonstrated by Wallach, has $[a]_{D\,22°}$ $-42,07°$. If this oxime is converted into the benzoyl compound, two modifications can be obtained by fractional crystallisation from ligroin, of which the one rotates to the right, and the other to the left.

R-d-α-benzoylcyclomethylhexanone oxime (melting point 96° to 97°) is the principal product, and has in ethereal solution $[a]_{D\,22°} + 19,97°$. The more readily soluble R-l-β-benzoylcyclomethylhexanone oxime (melting point 82° to 83°) has in ethereal solution $[a]_{D\,21°} - 86,08°$.

The fact that from one completely uniform oxime two benzoyl compounds of different rotations are obtained, can only be explained, as Wallach points out, by stereoisomerism of the nitrogenous radical, so that the two compounds must be interpreted as syn- and antiform of the benzoylised oxime, and in this manner the optical behaviour of the molecule is influenced one way or the other.

---

[1]) Bull. Soc. chim. III. **31** (1904), 899.
[2]) Liebig's Annalen **332** (1804), 337.

As 1- 1, 3-cyclomethylhexanone is up to the present unknown, it has not been possible to produce the corresponding bodies. From inactive 1, 3-cyclomethylhexanone prepared according to Knoevenagel[1]), a non-crystallising oxime was obtained, which, like the R-l-oxime, yielded two benzoyl compounds:

i-α-benzoylcyclomethylhexanone oxime of the melting point 105° to 106°,
i-β-benzoylcyclomethylhexanone oxime of the melting point 70° to 72°.

It has not yet been possible to obtain the expected similar results with methylpentanone, of analogous construction to methylhexanone, nor with menthone oxime, the isopropyl derivative of methylhexanone oxime.

Methylcyclohexenone. Paul Rabe has accomplished with the same results as with carvone the addition of aceto-acetic ester to methylcyclohexenone. The bicyclic ketone-alcohol 1-methyldicyclo-1, 3, 3-nonene-5-ol-7-one is a viscid oil of the boiling point 170° to 173° (17 to 18 mm. pressure). The glycol obtained from it by reduction, leaflets of the melting point 124° to 125°, when heated with hydriodic acid, yields the hydrocarbon 1-methyldicyclo-1, 3, 3-nonane, a colourless liquid with a terpene-like odour, of the boiling point 176° to 178° (751 mm. pressure).

Isoximes. As a continuation of his earlier experiments on isoximes, Wallach[2]) publishes a work in which he makes us acquainted with two new bodies of this description. The oximes of dihydroisophorone [methyl-(1)-dimethyl-(3)-cyclohexanone (5)], melting point 84° to 85°, and of methyl-(1)-dimethyl-(3)-cyclohexanone (6) yield on conversion each two isomeric isoximes with heterocyclic hepta-ring.

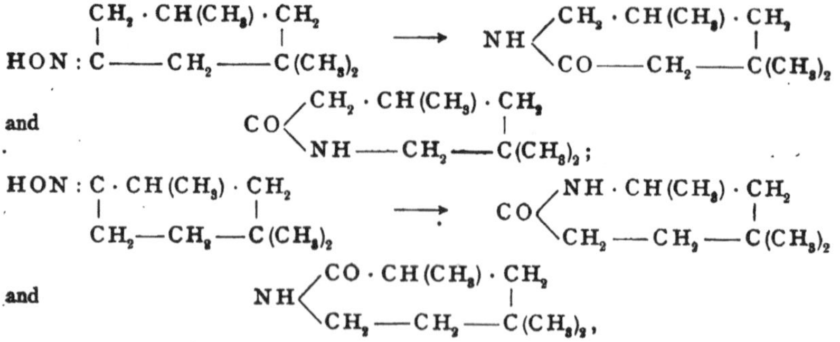

$$CH_2 \cdot CH(CH_3) \cdot CH_2$$
$$HON : C \text{——} CH_2 \text{——} C(CH_3)_2 \longrightarrow NH \Big\langle \begin{array}{l} CH_2 \cdot CH(CH_3) \cdot CH_2 \\ CO \text{——} CH_2 \text{——} C(CH_3)_2 \end{array}$$

and
$$CO \Big\langle \begin{array}{l} CH_2 \cdot CH(CH_3) \cdot CH_2 \\ NH \text{——} CH_2 \text{——} C(CH_3)_2 \end{array} ;$$

$$HON : C \cdot CH(CH_3) \cdot CH_2$$
$$CH_2 \text{—} CH_2 \text{——} C(CH_3)_2 \longrightarrow CO \Big\langle \begin{array}{l} NH \cdot CH(CH_3) \cdot CH_2 \\ CH_2 \text{——} CH_2 \text{——} C(CH_3)_2 \end{array}$$

and
$$NH \Big\langle \begin{array}{l} CO \cdot CH(CH_3) \cdot CH_2 \\ CH_2 \text{——} CH_2 \text{——} C(CH_3)_2 \end{array} ,$$

which differ from each other in the melting point and solubility.

---

[1]) Liebig's Annalen **297** (1897), 154.
[2]) Nachr. k. Ges. Wiss. Göttingen **1904**, 15. Abstr. according to Chem. Centralbl. **1904**, II, 653.

Dihydroisophorone yields an α-isoxime of the melting point 111°
to 112° and a β-isoxime of the melting point 82° to 84°. The
methyl-(1)-dimethyl-(3)-cyclohexanone-(6)-oxime melting at 108° to 109°,
when converted with concentrated sulphuric acid, yields an α-isoxime
of the melting point 115° to 116°, and a readily soluble β-isoxime
of the melting point 106° to 108°.

Phenols and phenol ethers.

Some few years ago we reported on a method worked out by
S. B. Schryver[1]), for determining phenols in essential oils, according
to which sodium amide reacts on phenols in such manner, that the
hydrogen of the phenol-group is replaced by sodium, with simultaneous
formation of ammonia. The latter is combined with acid and determined by titration. Owing to the great reaction-capacity of the
amide with water, the method is of course only applicable in the case
of absolutely dry mixtures and essential oils. Control-experiments
made at the time in our laboratories had given satisfactory results,
and we therefore intended making a more extensive use of this method
if the occasion should arise. We also expressed at the time the opinion
that this method might possibly become specially important for the
quantitative estimation of alcohols, particularly terpene alcohols (linalool,
terpineol). In the meantime we have made experiments in this direction,
but unfortunately with the result that the method is unserviceable for
the determination of alcohols, even for those of the aromatic series,
such as benzyl alcohol, etc. In every analysis carried out according
to the directions given, the values obtained were much too high, which
may possibly be explained by the fact that the sodium amide present
in excess in every determination, not only acts in the expected manner,
that is to say with formation of one molecule ammonia, but also
produces more far-reaching changes in the alcohol molecule, with formation
of more ammonia. Now as, according to Schryver's observations,
the method can only be applied to oils containing phenols, which are
free from ketones and aldehydes, the application of the method in
view of our observation with alcohols, would be limited solely to those
alcohols which represent mixtures of phenols with terpenes. As a
matter of fact, we have obtained satisfactory results in all such cases,
as for example with clove oil, pimenta oil, and artificial mixtures. On
the other hand, no useful values were obtained in determinations with
oil of thyme, where alcohols, such as linalool and borneol, are also
present.

---

[1]) Report October **1899**, 60; Journ. Soc. chem. Ind. **18** (1899), No. 6.

In the following table we give the determinations made by us, with the values obtained.

Substance	Corresponding quantity of phenol, or alcohol	cc. $\frac{n}{2}H_2SO_4$ used up for neutralising NH_3	Quantity of substance found	Difference	Difference in per cent.
		cc.	gm.	gm.	
Mixture of eugenol and turpentine oil (50%)	1,290 gm. eugenol	15,90	1,3038	0,0138	+ 0,54
Mixture of eugenol (75%) and turpentine oil (25%)	1,6125 gm. ,,	19,60	1,6072	0,0053	+ 0,25
Clove oil, about 95% eugenol (according to Umney-Schimmel & Co.)	0,9500 gm. ,, 0,8012 gm. ,,	11,50 9,75	0,9430 0,7995	0,0070 0,0017	— 0,70 — 0,20
Pimenta oil, 72,9% eugenol (according to Umney-Schimmel & Co.)	0,8529 gm. ,, 0,8857 gm. ,, 0,8529 gm. ,,	10,50 11,00 10,30	0,8610 0,9020 0,8446	0,0081 0,0163 0,0083	+ 0,69 + 1,34 — 0,72
Thyme oil, about 61% carvacrol (according to Umney-Schimmel & Co.)	0,7787 gm. carvacrol	14,35	1,6114	0,8327	+ 65,65
Thymol	1,0100 gm.	13,25	0,9937	0,0163	— 1,62
Carvacrol	1,5900 gm.	20,25	1,5199	0,0701	— 4,41
Geraniol	1,0028 gm. 1,0021 gm. 1,1079 gm.	25,40 19,20 25,00	1,9558 1,4784 1,9250	0,9530 0,4763 0,8171	+ 95,03 + 47,53 + 73,75
Terpineol	1,0156 gm. 1,4331 gm.	25,90 34,15	1,9943 2,6290	0,9787 1,1959	+ 96,36 + 83,45
Linalool	1,0100 gm. 1,2300 gm.	20,60 26,05	1,5862 2,0058	0,5762 0,7758	+ 57,05 + 63,07
Borneol	1,0550 gm.	22,30	1,7171	0,6621	+ 62,76
Menthol	1,0200 gm.	21,40	1,6692	0,6492	+ 63,60
Benzyl alcohol	1,0600 gm.	24,55	1,3257	0,2657	+ 25,07
Phenylpropyl alcohol	1,0000 gm.	24,85	1,6896	0,6896	+ 68,96

It has until recently been a generally accepted proposition (which has also found its way into most text-books) that in the zinc dust distillation of phenol ethers and their derivatives, the alkoxy-group remains unchanged, although 18 years ago E. Bamberger[1] demonstrated that with increased temperature and high pressure phenol ethers are decomposed into phenols and ethylene hydrocarbons according to the equation $C_6H_5 \cdot OC_nH_{2n+1} = C_6H_5OH + C_nH_{2n}$. Various divergent results which Thoms[2] obtained in the zinc dust distillation of higher phenol ethers induced him to test this reaction with the most simple

[1] Berl. Berichte 19 (1886), 1818.
[2] Archiv der Pharm. 242 (1904), 95.

body of this class, viz., anisol. In 24 distillations he obtained from 60 gm. anisol 20 gm. distillate, in which were found about 9 gm. unconverted anisol, about 3 gm. benzene, 3 gm. phenol, and 2,5 gm. diphenyl. Of the volatile products of distillation, benzene was retained as p-dibrombenzene and ethylene as dibromide, by means of a receiver charged with bromine. It follows that the stability of phenol ethers in the distillation with zinc dust is but slight.

After a short historical review of the publications dealing with nitrosocarvacrol, E. Kremers and J. W. Brandel[1]) devote a detailed study to the various methods now in use for producing this preparation. During a series of experiments made with a view to determining the molecular weights of nitrosocarvacrol and nitrosothymol by the boiling point method, Kremers and Brandel found that the molecular weight ascertained gives different values according to the nature of the solvent (alcohol, chloroform, or benzene) and the degree of concentration of the solutions. From the results the general conclusion may apparently be drawn, that in the case of the nitroso-derivatives of thymol and carvacrol a tendency exists to form double molecules in solution, but that such a tendency is only in one instance approximately satisfied, namely in the case of benzene.

By adding silver nitrate to the neutral solution of nitrosocarvacrol in sodium hydroxide, Kremers and Brandel[2]) obtained two silver compounds of nitrosocarvacrol, $C_{10}H_{12}NOAg$ and $C_{10}H_{11}NOAg_2$.

By allowing nitrous acid to act on myristicin, E. Rimini[3]) endeavours to bring further support to the view (already confirmed by Thoms) that myristicin contains an allyl side-chain. The nitrosite formed in the first instance, $C_8H_7O_3C_3H_5 \cdot N_2O_3$, is converted by treatment with absolute alcohol into a nitro-oxime $C_8H_7O_3 \cdot CH_2 \cdot C(NOH) \cdot CH_2NO_2$, from which the nitroketone is regenerated by boiling with sulphuric acid. The presence of the nitro-group is proved by the capacity of being reduced into a ketone-amine which is identified by the picrate. When treated with mercuric acetate according to Balbiano's method, myristicin yields the following compound:

$$CH_2O_2 \cdot (OCH_3)C_6H_2 \cdot C_3H_5 \Big\langle {}^{HgC_2H_3O_2}_{OH}$$

In the opinion of the author, this compound is very suitable for the production of myristicin in the pure state.

---

[1]) Pharmaceut. Review **22** (1904), 248.
[2]) Pharmaceut. Review **22** (1904), 290.
[3]) Rendiconti della Società Chimica di Roma **2** (1904), 20.

Since for parsley apiol the formula of a 1-allyl-2,5-dimethoxy-3,4-methylene dioxybenzene had been established[1]), there could only come under consideration for dill apiol which Thoms had found in matico oil 1-allyl-5,6-dimethoxy-3,4-methylene dioxybenzene, of which the correctness has been demonstrated by Thoms[2]). The argumentation follows closely on the lines of the one previously given for parsley apiol, and, like the latter, culminates in the displacement of two alkoxy-groups in the para-position, with formation of a quinone.

Conjointly with A. Biltz, Thoms[3]) publishes a communication on the relationship between safrol, eugenol, and asarone. From dihydrosafrol (obtained by reduction from isosafrol with sodium and alcohol), there could be produced by treatment with glacial acetic and nitric acid, a nitro-compound of the formula

$$\text{C}_3\text{H}_7 \left\langle \begin{array}{c} \text{H} \quad\quad \text{O} - \text{CH}_2 \\ \\ \text{NO}_2 \quad\quad \text{H} \end{array} \right\rangle \text{O}$$

of the melting point $36°$, whose methylenedioxy-group could be split up with aluminium chloride. By etherifying with dimethylsulphate the mixture of isomeric phenols produced, Thoms and Biltz obtained a nitrodihydromethyl eugenol of the melting point $81°$, which was identic with the one formed from isoeugenol. On reduction with aluminium amalgam, the nitrodihydrosafrol was converted into the little stable corresponding amino compound of the melting point $24°$; when nitrated further, the nitro-product yielded a dinitrodihydrosafrol from which a diamino body was produced by means of aluminium amalgam; with ammonium sulphide a nitroamino-compound was formed in which the second nitro-group was reduced, as after elimination of the amino-group the original product of the melting point $36°$ was again recovered. On saponification of the nitrodihydromethyl-eugenol with aluminium chloride there were formed two isomeric nitrophenols of the melting points $52°$ and $78°$; the former was ethylated with ethyl iodide, the ether reduced into the amine, and the latter oxidised by means of bichromate mixture into the quinone

$$\text{C}_3\text{H}_7 \left\langle \begin{array}{c} \text{H} \quad\quad \text{O} \\ \\ \text{O} \quad\quad \text{H} \end{array} \right\rangle \text{O}\,\text{CH}_3$$

[1]) Berl. Berichte **36** (1903), 1714. — Report October **1903**, 99.
[2]) Arch. der Pharm. **242** (1904), 344.
[3]) Arch. der Pharm. **242** (1904), 85.

which was found to be identic with the quinone obtained from asarone (see below) of the melting point 111°.

This quinone obtained by Thoms and Biltz from nitrodihydro methyleugenol, was obtained by Beckstroem[1]) from asarone by oxidation of its dihydro derivative with chromic acid.

The last-named author also examined the behaviour of asarone dibromide

$$C_6H_2 \underset{CHBr \cdot CHBr \cdot CH_3}{\overset{(OCH_3)_3}{<}}$$

towards sodium methylate. With this reagent the dibromide yielded, in agreement with observations made by Auwers and Müller[2]), a brominated ether of the formula

$$C_6H_2 \underset{CH(OCH_3) \cdot CHBr \cdot CH_3}{\overset{(OCH_3)_3}{<}}$$

on the other hand, when the dibromide was treated with an excess of methylate, with application of heat, it did not yield the expected ketone

$$C_6H_2 \underset{CO \cdot CH_2 \cdot CH_3,}{\overset{(OCH_3)_3}{<}}$$

but a high-boiling oil, from which, after prolonged standing, a very small quantity of crystals melting at 106°, not sufficient for analysis, was separated off.

The great facility with which the pure asarone dibromide melting at 86° becomes decomposed is very striking; even in the vacuum desiccator it acquired a dark colour after a few weeks, and separated off a body melting at 109°. Beckstroem concludes from the bromine-determination that a condensation of two molecules asarone with one atom bromine has taken place. Of a chiefly preparative interest are the condensations made with asaryl aldehyde, the oxidation product of asarone, and various bodies such as alcohol, acetone, and methyl-amylketone.

=====

Benzyl acetate. The production of this useful perfume on a large scale enabled us to reduce its price still further. In the hands of an intelligent expert, this body is very serviceable for purposes of perfumery, and we would once more call attention to the fact that it is an important constituent of natural jasmine and ylang ylang oils. This should be an inducement to give the article a thorough trial.

[1]) Arch. der Pharm. **242** (1904), 98.
[2]) Berl. Berichte **35** (1902), 114. — Report April **1902**, 95.

Borneol (Borneo camphor). Of this article our production was for a time fully occupied, so much so, that we were repeatedly compelled to abstain from making offers. The consumption has experienced such an enormous increase during the last few years in India and Siam, that it would hardly be possible to satisfy the demand, if simplifications in the manufacturing process had not assisted the production.

Citral. The high value of lemongrass oil also brought about high prices for this preparation. The abnormally hot summer has greatly influenced the consumption. The use of it has become more and more general since it has been found that the citral obtained from the above-named material is fully equal in taste to that produced from lemon oil, and that the disparagement with which it was at first treated, was due to prejudice. Our pure commercial citral is in every respect identical with that of lemon oil, and cannot be distinguished from the latter. There are, however, products met with in commerce which may throw discredit on the preparation, and it is impossible to caution consumers too strongly against such kinds.

Coumarin. Of this product the most diverse qualities are found on the market, and for this reason a direct comparison of the prices it not permissible. Only by quite special purification methods is it possible to obtain a product which in its odour is absolutely equal to the coumarin from the tonka-bean, and a product of this kind cannot be sold at the lowest cut prices. We have come across a coumarin which had in a pronounced manner the odour of the crude material, carbolic acid. The important increase in the prices of all raw materials has, unfortunately, had no influence on the selling price of coumarin.

Eman. Senft[1] makes some communications on the presence and the detection of coumarin in the tonka bean. According to these, the pods of *Dipterix odorata* Willd. do not contain coumarin; on the contrary, it is found in the tissue cells of the cotyledons dissolved in the fatty oil. The coumarin-content of the beans may be as high as 10%. According to Vogel, the separation of coumarin is due to the fact that in consequence of the shrinking of the peripheral cells of the perisperm, the fatty oil is first of all squeezed out, and the coumarin is then deposited from the latter on to the cotyledons turned towards each other, either below or on the testa. — With iodine, coumarin forms a crystalline compound; the smallest quantities of solid coumarin, or even of coumarin dissolved in water, can be detected by means of zinc chloride and iodine.

---

[1] Apotheker Zeitg. **19** (1904), 271.

Eucalyptol. The demand for this pure body was as important as it was regular, but could be satisfied completely, as the market offered a sufficient selection of oils rich in eucalyptol, at moderate prices. A change in the prices is not to be expected in the near future.

H. Thoms and B. Molle have published a treatise on the reduction-products of eucalyptol[1]), of which the principal points have already been dealt with in our last Report[2]).

Geraniol. The present price of citronella oil would justify an increase in the quotations, but improved manufacturing methods enable us to abstain from this for the present.

Heliotropin. With regard to this important article, nothing new can be reported. For good qualities, such as our product, the present prices are willingly paid; when properly calculated they leave only a very modest profit.

Our extra-purified quality is very much in demand.

Menthol, recryst. pur. During the last six months sales have again been very important. The brisk demand, favoured by declining prices, still continues, and the present low value is also looked upon as suitable for speculative purchases, on the assumption that the war in the Far East is bound to affect this article. This view has been so wide-spread and firmly rooted ever since the commencement of the war, that we had to use every argument to restrain clients from making large contracts at a time when we had already received information as to the important results of this year's peppermint-harvest in Japan.

To the last-named cause alone the present low prices are due, but they appear to have now reached their lowest level. A favourable influence is exerted on this article by the failure of the peppermint-harvest in America, and by the constantly growing consumption of menthol, which is now used on a large scale for cosmetic-therapeutic remedies.

Statistics respecting the shipments of crude menthol during the first six months of this year are found on pp. 75/76 of the present Report.

For the purpose of obtaining clear aqueous solutions of menthol[3]), sarsaparilla tincture has lately come into use in the place of quillaya tincture which on account of its content of saponin and glucoside is

---

[1]) Arch. der Pharm. **242** (1904), 181.
[2]) Report April **1904**, 60.
[3]) Pharm. Centralh. **45** (1904), 180.

CPSIA information can be obtained
at www.ICGtesting.com
Printed in the USA
BVHW070822040219
539400BV00034B/2401/P